UTILITY MANAGEMENT

Second Edition

A Field Study Training Program

Prepared by

Lorene Lindsay

for

Office of Water Programs
College of Engineering and Computer Science
California State University, Sacramento

2004

In recognition of the need to preserve natural resources, this manual is printed using recycled paper. The text paper and the cover are composed of 10% post-consumer waste. The Office of Water Programs strives to increase its commitment to sustainable printing practices.

Funding for this operator training manual was provided by the Office of Water Programs, California State University, Sacramento. Mention of trade names or commercial products does not constitute endorsement or recommendation for use by the Office of Water Programs or California State University, Sacramento.

Copyright © 2004 by
University Enterprises, Inc.

First edition published 1998.

All Rights Reserved.
Printed in the United States of America

24 23 22 21 20   4 5 6 7 8

ISBN
1-884701-49-3

www.owp.csus.edu

OPERATOR TRAINING MATERIALS

OPERATOR TRAINING MANUALS AND VIDEOS IN THIS SERIES are available from the Office of Water Programs, California State University, Sacramento, 6000 J Street, Sacramento, CA 95819-6025, phone: (916) 278-6142, e-mail: wateroffice@csus.edu, FAX: (916) 278-5959, or website: www.owp.csus.edu.

1. *UTILITY MANAGEMENT*,
2. *MANAGE FOR SUCCESS*,
3. *OPERATION OF WASTEWATER TREATMENT PLANTS*, 2 Volumes,
4. *ADVANCED WASTE TREATMENT*,
5. *INDUSTRIAL WASTE TREATMENT*, 2 Volumes,
6. *PRETREATMENT FACILITY INSPECTION,**
7. *OPERATION AND MAINTENANCE OF WASTEWATER COLLECTION SYSTEMS*, 2 Volumes,* (Spanish edition available),
8. *TREATMENT OF METAL WASTESTREAMS*,
9. *SMALL WASTEWATER SYSTEM OPERATION AND MAINTENANCE*, 2 Volumes,
10. *WATER TREATMENT PLANT OPERATION*, 2 Volumes,
11. *SMALL WATER SYSTEM OPERATION AND MAINTENANCE,** and
12. *WATER DISTRIBUTION SYSTEM OPERATION AND MAINTENANCE*.

Other materials and training aids developed by the Office and Water Programs to assist operators in improving the operation and maintenance and overall performance of their systems include:

- *COLLECTION SYSTEMS: METHODS FOR EVALUATING AND IMPROVING PERFORMANCE.* This handbook presents detailed benchmarking procedures and worksheets for using performance indicators to evaluate the adequacy and effectiveness of existing O & M programs. It also describes how to identify problems and suggests many methods for improving the performance of a collection system.

- *OPERATION AND MAINTENANCE TRAINING VIDEOS.* This series of six 30-minute videos demonstrates the equipment and procedures collection system crews use to safely and effectively operate and maintain their collection systems. These videos complement and reinforce the information presented in Volumes I and II of *OPERATION AND MAINTENANCE OF WASTEWATER COLLECTION SYSTEMS*.

- *SMALL WATER SYSTEMS VIDEO INFORMATION SERIES.* This series of ten training videos was prepared for the operators, managers, owners, and board members of small drinking water systems. Topics covered include operators' roles and responsibilities; safety, operation, and maintenance of surface and groundwater treatment, distribution, and storage facilities; monitoring; administration; financial management; and also emergency preparedness and response.

- *PRETREATMENT FACILITY INSPECTION TRAINING VIDEOS.* This series of five 30-minute videos demonstrates the procedures to effectively inspect an industry, measure flows, and collect samples. These videos complement and reinforce the information presented in *PRETREATMENT FACILITY INSPECTION*.

The Office of Water Programs at California State University, Sacramento, has been designated by the U.S. Environmental Protection Agency as a *SMALL PUBLIC WATER SYSTEMS TECHNOLOGY ASSISTANCE CENTER*. This recognition will provide funding for the development of training videos for the operators and managers of small public water systems. Additional training materials will be produced to assist the operators and managers of small systems.

PREFACE TO UTILITY MANAGEMENT

Operators have expressed a need for more information on utility management for many years. The Office of Water Programs at California State University, Sacramento, was aware of this need for training. Operation and maintenance is the major area of expertise of the Office of Water Programs. Many operators successfully completed our training programs, advanced in our profession, and became utility managers. These operators urged the Office of Water Programs to develop training programs for utility managers similar to the operator O & M training programs.

The need for a utility management training program prompted several qualified utility management trainers to approach the Office of Water Programs and offer to prepare a utility management training program. Unfortunately several early attempts did not produce training materials meeting the standards of the Office of Water Programs.

Lorene Lindsay, a respected and successful trainer in the field of utility management training, offered to prepare utility management training material for the Office of Water Programs. The Association of Boards of Certification provided validated "Need to Know" criteria for utility managers, supervisors, and administrators. Ms. Lindsay prepared training material covering the topics of utility planning, organizing, staffing, communicating, conducting meetings, public relations, and financial management. California Water Environment Association (CWEA) members Lynne Scarpa, Phil Scott, Chris Smith, and Rich von Langen reviewed the material for technical accuracy and also made many valuable suggestions for more effective presentation of management concepts and issues. Office of Water Programs staff members Peg Hannah and Janet Weeks incorporated the reviewers' comments and arranged the information in the self-study or home-study format.

The original intent was to add a utility management chapter to *OPERATION OF WASTEWATER TREATMENT PLANTS*, Volume II, and *WATER TREATMENT PLANT OPERATION*, Volume II. This will be done. The Office of Water Programs attempted to determine the best approach to include utility management training materials in our other training programs, such as wastewater collection systems, water distribution systems, small water and wastewater systems, industrial waste treatment, and pretreatment inspection. Also there are many operators who have completed our training programs who now need a training program on utility management. For these reasons the Office of Water Programs decided to prepare this separate training manual dedicated to the needs of utility managers.

I wish to take this opportunity to thank the many operators who suggested we prepare this training manual and also suggested important topics that should be covered. We expect this training manual will help you continue to advance in our profession.

Whenever California State University, Sacramento, reprints one of the operator training manuals, the material is updated in accordance with the comments and suggestions received from the operators enrolling in courses which use the manual. While you are reading the material in this manual, please make notes of questions and areas where you would improve the material. By sending your comments and suggestions to me, operators who use this manual in the future will benefit from your experience, knowledge, and contributions.

1998

Kenneth D. Kerri
Office of Water Programs
California State University, Sacramento
6000 J Street
Sacramento, CA 95819-6025
Phone: (916) 278-6142

UTILITY MANAGEMENT
OBJECTIVES

Following completion of this manual, you should be able to:

1. Identify the functions of a manager,

2. Describe the benefits of short-term, long-term, and emergency planning,

3. Define the following terms:
   - a. Authority,
   - b. Responsibility,
   - c. Delegation,
   - d. Accountability, and
   - e. Unity of command,

4. Read and construct an organizational chart identifying lines of authority and responsibility,

5. Write a job description for a specific position within the utility,

6. Write good interview questions,

7. Conduct employee evaluations,

8. Describe the steps necessary to provide equal and fair treatment to all employees,

9. Prepare a written or oral report on the utility's operations,

10. Communicate effectively within the organization, with media representatives, and with the community,

11. Describe the financial strength of your utility,

12. Calculate your utility's operating ratio, coverage ratio, and simple payback,

13. Prepare a contigency plan for emergencies,

14. Prepare a plan to strengthen the security of your utility's facilities,

15. Set up a safety program for your utility, and

16. Collect, organize, file, retrieve, use, and dispose of utility records.

UTILITY MANAGEMENT
USES OF THIS MANUAL

This manual was developed to serve the needs of utility managers in several different situations. The format was developed to serve as a home-study or self-paced instruction course for managers in remote areas or persons unable to attend formal classes either due to shift work, personal reasons, or the unavailability of suitable classes. This home-study training program uses the concept of self-paced instruction where you are your own instructor and work at your own speed. In order to certify that a person has successfully completed this program, an objective test and special answer sheet are provided when a person enrolls in this course.

Also, this manual can serve effectively as a textbook in the classroom. Many colleges and universities have used this manual as a text in formal classes (often taught by managers). In areas where colleges are not available or are unable to offer classes in utility management, managers can join together to offer their own courses using this manual.

Utilities can use the manual in several types of on-the-job training programs. In one type of program, a manual is purchased for each manager. A senior manager or a group of managers are designated as instructors. These managers help answer questions when the persons in the training program have questions or need assistance. The instructors grade the objective test at the end, record scores, and notify California State University, Sacramento, of the scores when a person successfully completes this program. This approach eliminates any waiting while papers are being graded and returned by CSUS. This manual was prepared to help managers operate and maintain their utilities. Please feel free to use the manual in the manner which best fits your training needs and the needs of other managers. We will be happy to assist you in developing your training program. Please feel free to contact:

Program Director
Office of Water Programs
California State University, Sacramento
6000 J Street
Sacramento, California 95819-6025
Phone: (916) 278-6142

INSTRUCTIONS TO PARTICIPANTS IN HOME-STUDY COURSE

Procedures for reading the lessons and answering the questions are contained in this section.

To progress steadily through this program, you should establish a regular study schedule. For example, many managers in the past have set aside two hours during two evenings a week for study.

The study material is contained in two lessons. The time required to complete each lesson will depend on your background and experience. Some people might require 1 hour to complete a lesson and some might require 3 hours; but that is perfectly all right. *THE IMPORTANT THING IS THAT YOU UNDERSTAND THE MATERIAL IN THE LESSON!*

Each lesson is arranged for you to read a few paragraphs, write the answers to the questions at the end of each section, check your answers against suggested answers; and then *YOU* decide if you understand the material sufficiently to continue or whether you should read the paragraphs again. You will find that this procedure is slower than reading a normal textbook, but you will remember much more when you have finished the lesson.

Some discussion and review questions are provided following each lesson. These questions review the important points you have covered in the lesson. Write the answers to these questions in your notebook.

After you have completed Section 15, you will find a final examination. This exam is provided for you to review how well you remember the material. You may wish to review the entire manual before you take the final exam. Some of the questions are essay-type questions which are used by some states for higher level certification examinations. After you have completed the final examination, grade your own paper and determine the areas in which you might need additional review before your next certification or civil service examination.

You are your own teacher in this program. You could merely look up the suggested answers from the answer sheet or copy them from someone else, but you would not understand the material. Consequently, you would not be able to apply the material to the management of your facility nor recall it during an examination for certification or a civil service position.

YOU WILL GET OUT OF THIS PROGRAM WHAT YOU PUT INTO IT.

SUMMARY OF PROCEDURE

A. OPERATOR (YOU)

1. Read what you are expected to learn (the objectives).
2. Read sections in the lesson.
3. Write your answers to questions at the end of each section in your notebook. You should write the answers to the questions just as you would if these were questions on a test.
4. Check your answers with the suggested answers which begin on page 50.
5. Decide whether to reread the section or to continue with the next section.
6. Write your answers to the discussion and review questions at the end of each lesson in your notebook.

ENROLLMENT FOR CREDIT AND CERTIFICATE

Students wishing to earn credits and a certificate for completing this course may enroll by contacting the Office of Water Programs, California State University, Sacramento, 6000 J Street, Sacramento, CA 95819-6025, (916) 278-6142. If you have already enrolled, the enrollment packet you were sent contains detailed instructions for completing and returning the objective tests. Please read these important instructions carefully before marking your answer sheets.

Following successful completion of each volume in this program, a Certificate of Completion will be sent to you. If you wish, the Certificate can be sent to your supervisor, the mayor of your town, or any other official you think appropriate. Some operators have been presented their Certificate at a City Council meeting, got their picture in the newspaper, and received a pay raise.

UTILITY MANAGEMENT
TABLE OF CONTENTS

	Page
OBJECTIVES	v
USES OF THIS MANUAL	vi
INSTRUCTIONS TO PARTICIPANTS IN HOME-STUDY COURSE	vi
SUMMARY OF PROCEDURE	vii
ENROLLMENT FOR CREDIT AND CERTIFICATE	vii
UTILITY MANAGEMENT WORDS	xi

LESSON 1

			Page
1	NEED FOR UTILITY MANAGEMENT		1
2	FUNCTIONS OF A MANAGER		1
3	PLANNING		3
4	ORGANIZING		3
5	STAFFING		6
	5.0	The Utility Manager's Responsibilities	6
	5.1	How Many Employees Are Needed?	7
	5.2	Qualifications Profile	7
	5.3	Applications and the Selection Process	8
		5.30 Advertising the Position	8
		5.31 Paper Screening	8
		5.32 Interviewing Applicants	8
		5.33 Selecting the Most Qualified Candidate	11
	5.4	New Employee Orientation	11
	5.5	Employment Policies and Procedures	11
		5.50 Probationary Period	11
		5.51 Compensation	12
		5.52 Training and Certification	12
		5.53 Performance Evaluation	13
		5.54 Dealing With Disciplinary Problems	13
		5.55 Example Policy: Harassment	17
		5.56 Laws Governing Employer/Employee Relations	20
		5.57 Personnel Records	21
	5.6	Unions	21

| | DISCUSSION AND REVIEW QUESTIONS | 22 |

LESSON 2

6	COMMUNICATION		23
	6.0	Oral Communication	23
	6.1	Written Communication	23
7	CONDUCTING MEETINGS		25
8	PUBLIC RELATIONS		26
	8.0	Establish Objectives	26
	8.1	Utility Operations	26
	8.2	The Mass Media	26
	8.3	Being Interviewed	26
	8.4	Public Speaking	27
	8.5	Telephone Contacts	27
	8.6	Customer Inquiries	27
	8.7	Plant Tours	28
9	FINANCIAL MANAGEMENT		28
	9.0	Financial Stability	28
	9.1	Budgeting	29
	9.2	Equipment Repair/Replacement Fund	30
	9.3	Capital Improvements and Funding in the Future	30
	9.4	Financial Assistance	31
10	OPERATIONS AND MAINTENANCE		31
	10.0	The Manager's Responsibilities	31
	10.1	Purpose of O & M Programs	32
	10.2	Types of Maintenance	32
	10.3	Benefits of Managing Maintenance	33
11	EMERGENCY RESPONSE PLAN		33
12	HOMELAND DEFENSE		34
13	SAFETY PROGRAM		39
	13.0	Policy Statement	39
	13.1	Responsibilities	39
	13.2	Hazard Communication Program and Worker Right-To-Know (RTK) Laws	39
	13.3	Confined Space Entry Procedures	43
	13.4	Reporting	43
14	RECORDKEEPING		47
	14.0	Purpose of Records	47
	14.1	Computer Recordkeeping Systems	47
	14.2	Types of Records	47
	14.3	Equipment and Maintenance Records	47

	14.4	Plant Operations Data	47
	14.5	Procurement Records	47
	14.6	Inventory Records	49
	14.7	Personnel Records	49
	14.8	Disposition of Plant and System Records	49
15	ACKNOWLEDGMENTS		49
16	ADDITIONAL READING		49
DISCUSSION AND REVIEW QUESTIONS			50
SUGGESTED ANSWERS			50
FINAL EXAMINATION			55
SUGGESTED ANSWERS FOR FINAL EXAMINATION			57
SUBJECT INDEX			59

UTILITY MANAGEMENT
WORDS

ACEOPS
See ALLIANCE OF CERTIFIED OPERATORS, LAB ANALYSTS, INSPECTORS, AND SPECIALISTS (ACEOPS).

ACCOUNTABILITY
When a manager gives power/responsibility to an employee, the employee ensures that the manager is informed of results or events.

ALLIANCE OF CERTIFIED OPERATORS, LAB ANALYSTS, INSPECTORS, AND SPECIALISTS (ACEOPS)
A professional organization for operators, lab analysts, inspectors, and specialists dedicated to improving professionalism; expanding training, certification, and job opportunities; increasing information exchange; and advocating the importance of certified operators, lab analysts, inspectors, and specialists. For information on membership, contact ACEOPS, 1810 Bel Air Drive, Ames, IA 50010-5125, phone (515) 663-4128 or e-mail: Info@aceops.org.

AUTHORITY
The power and resources to do a specific job or to get that job done.

BOND
(1) A written promise to pay a specified sum of money (called the face value) at a fixed time in the future (called the date of maturity). A bond also carries interest at a fixed rate, payable periodically. The difference between a note and a bond is that a bond usually runs for a longer period of time and requires greater formality. Utility agencies use bonds as a means of obtaining large amounts of money for capital improvements.

(2) A warranty by an underwriting organization, such as an insurance company, guaranteeing honesty, performance, or payment by a contractor.

CALL DATE
First date a bond can be paid off.

CERTIFICATION EXAMINATION
An examination administered by a state agency or professional association that managers take to indicate a level of professional competence. Operator certification is mandatory in the United States for the Chief Operators of water treatment plants, water distribution systems, and wastewater treatment plants.

CERTIFIED OPERATOR
A person who has the education and experience required to operate a specific class of treatment facility as indicated by possessing a certificate of professional competence given by a state agency or professional association.

CODE OF FEDERAL REGULATIONS (CFR)
A publication of the United States Government which contains all of the proposed and finalized federal regulations, including environmental regulations.

CONFINED SPACE
Confined space means a space that:

A. Is large enough and so configured that an employee can bodily enter and perform assigned work; and

B. Has limited or restricted means for entry or exit (for example, manholes, tanks, vessels, silos, storage bins, hoppers, vaults, and pits are spaces that may have limited means of entry); and

C. Is not designed for continuous employee occupancy.

(Definition from the Code of Federal Regulations (CFR) Title 29 Part 1910.146.)

CONFINED SPACE, PERMIT-REQUIRED (PERMIT SPACE)

A confined space that has one or more of the following characteristics:

- Contains or has a potential to contain a hazardous atmosphere,
- Contains a material that has the potential for engulfing an entrant,
- Has an internal configuration such that an entrant could be trapped or asphyxiated by inwardly converging walls or by a floor which slopes downward and tapers to a smaller cross section, or
- Contains any other recognized serious safety or health hazard.

(Definition from the Code of Federal Regulations (CFR) Title 29 Part 1910.146.)

COVERAGE RATIO

The coverage ratio is a measure of the ability of the utility to pay the principle and interest on loans and bonds (this is known as "debt service") in addition to any unexpected expenses.

DEBT SERVICE

The amount of money required annually to pay the (1) interest on outstanding debts; or (2) funds due on a maturing bonded debt or the redemption of bonds.

DELEGATION

The act in which power is given to another person in the organization to accomplish a specific job.

GIS

Geographic **I**nformation **S**ystem. A computer program that combines mapping with detailed information about the physical locations of structures such as pipes, valves, and manholes within geographic areas. The system is used to help operators and maintenance personnel locate utility system features or structures and to assist with the scheduling and performance of maintenance activities.

GEOGRAPHIC INFORMATION SYSTEM (GIS)

A computer program that combines mapping with detailed information about the physical locations of structures such as pipes, valves, and manholes within geographic areas. The system is used to help operators and maintenance personnel locate utility system features or structures and to assist with the scheduling and performance of maintenance activities.

MSDS

Material **S**afety **D**ata **S**heet. A document which provides pertinent information and a profile of a particular hazardous substance or mixture. An MSDS is normally developed by the manufacturer or formulator of the hazardous substance or mixture. The MSDS is required to be made available to employees and operators whenever there is the likelihood of the hazardous substance or mixture being introduced into the workplace. Some manufacturers are preparing MSDSs for products that are not considered to be hazardous to show that the product or substance is *NOT* hazardous.

OSHA (O-shuh)

The Williams-Steiger **O**ccupational **S**afety and **H**ealth **A**ct of 1970 (OSHA) is a federal law designed to protect the health and safety of industrial workers, including the operators of water supply and treatment systems and wastewater treatment plants. The Act regulates the design, construction, operation, and maintenance of water supply systems, water treatment plants, wastewater collection systems, and wastewater treatment plants. OSHA also refers to the federal and state agencies which administer the OSHA regulations.

OPERATING RATIO

The operating ratio is a measure of the total revenues divided by the total operating expenses.

ORGANIZING

Deciding who does what work and delegating authority to the appropriate persons. A utility should have a written organizational plan and written policies.

OUCH PRINCIPLE

This principle says that as a manager when you delegate job tasks you must be **O**bjective, **U**niform in your treatment of employees, **C**onsistent with utility policies, and **H**ave job relatedness.

PLANNING

Management of utilities to build the resources and financial capability to provide for future needs.

PRESENT WORTH

The value of a long-term project expressed in today's dollars. Present worth is calculated by converting (discounting) all future benefits and costs over the life of the project to a single economic value at the start of the project. Calculating the present worth of alternative projects makes it possible to compare them and select the one with the largest positive (beneficial) present worth or minimum present cost.

RESPONSIBILITY

Answering to those above in the chain of command to explain how and why you have used your authority.

UTILITY MANAGEMENT

(Lesson 1 of 2 Lessons)

1 NEED FOR UTILITY MANAGEMENT

The management of a public or private utility, large or small, is a complex and challenging job. Communities are concerned about their drinking water and their wastewater. They are aware of past environmental disasters and they want to protect their communities, but they want this protection with a minimum investment of money. In addition to the local community demands, the utility manager must also keep up with increasingly stringent regulations and monitoring from regulatory agencies. While meeting these "external" (outside the utility) concerns, the manager faces the normal challenges from within the organization: personnel, resources, equipment, and preparing for the future. For the successful manager, all of these responsibilities combine to create an exciting and rewarding job.

A brief quiz is given in Table 1 that asks some basic management questions. This quiz can be used as a guide to management areas that may need some attention in your utility. You should be able to answer yes to most of the questions; however, all utilities have areas which can be improved.

In the environmental field, as well as other fields, the workforce itself is changing. Minorities, women, and people with disabilities provide new opportunities for growth in the utility. For the employee, however, overcoming employment barriers can be difficult, especially when the workload is demanding and physically challenging. The utility manager must provide adequate support services for these workers and learn to deal with organized worker groups.

Changes in the environmental workplace also are created by advances in technology. The environmental field has exploded with new technologies, such as video monitoring, robotics in the collection system, and computer-controlled treatment processes. The utility manager must keep up with these changes and provide the leadership to keep everyone at the utility up to speed on new ways of doing things. In addition, the utility manager must provide a safer, cleaner work environment while constantly training and retraining employees to understand new technologies.

QUESTIONS

Write your answers in a notebook and then compare your answers with those on page 50. By writing down the answers to these questions you will be helping yourself to remember important information in this manual.

1.0A What are the local community demands on a utility manager?

1.0B What has created changes in the environmental workplace?

2 FUNCTIONS OF A MANAGER

The functions of a utility manager are the same as for the CEO (Chief Executive Officer) of any big company: planning, organizing, staffing, directing, and controlling. In many small communities the utility manager may be the only one who has these responsibilities and the community depends on the manager to handle everything.

Planning (see Section 3) consists of determining the goals, policies, procedures, and other elements to achieve the goals and objectives of the agency. Planning requires the manager to collect and analyze data, consider alternatives, and then make decisions. Planning must be done before the other managing functions. Planning may be the most difficult in smaller communities, where the future may involve a decline in population instead of growth.

Organizing (see Section 4) means that the manager decides who does what work and delegates authority to the appropriate operators. The organizational function in some utilities may be fairly loose while some communities are very tightly controlled.

Staffing (see Section 5) is the recruiting of new operators and staff and determining if there are enough qualified operators and staff to fill available positions. The utility manager's staffing responsibilities include selecting and training employees, evaluating their performance, and providing opportunities for advancement for operators and staff in the agency.

Directing includes guiding, teaching, motivating, and supervising operators and utility staff members. Direction includes issuing orders and instructions so that activities at the facilities or in the field are performed safely and are properly completed.

TABLE 1 HOW WELL DOES YOUR SYSTEM MANAGE?

The following self-test is designed for small water or wastewater treatment facilities to provide a guide for identifying areas of concern and for improving small system management.

1. Is the treatment system budget separate from other accounts so that the true cost of treatment can be determined?
2. Are the funds adequate to cover operating costs, debt service, and future capital improvements?
3. Do operational personnel have input into the budget process?
4. Is there a monthly or quarterly review of the actual operating costs compared to the budgeted costs?
5. Does the user charge system adequately reflect the cost of treatment?
6. Are all users properly metered and does the unaccounted for water not exceed 20 percent of the total flow?
7. Are plant discharges and monitoring tests representative of plant performance?
8. Are operational control decisions based on process control testing within the plant?
9. Are provisions made for continued training for plant personnel?
10. Are qualified personnel available to fill job vacancies and is job turnover relatively low?
11. Are the energy costs for the system not more than 20 to 30 percent of the total operating costs?
12. Is the ratio of corrective (reactive) maintenance to preventive (proactive) maintenance remaining stable and is it less than 1.0?
13. Are maintenance records available for review?
14. Is the spare parts inventory adequate to prevent long delays in equipment repairs?
15. Are old or outdated pieces of equipment replaced as necessary to prevent excessive equipment downtime, inefficient process performance, or unreliability?
16. Are technical resources and tools available for repairing, maintaining, and installing equipment?
17. Is the utility's equipment providing the expected design performance?
18. Are standby units for key equipment available to maintain process performance during breakdowns or during preventive maintenance activities?
19. Are the plant processes adequate to meet the demand for treatment?
20. Does the facility have an adequate emergency response plan including a wastewater bypass storage procedure or an alternate water source?

Controlling involves taking the steps necessary to ensure that essential activities are performed so that objectives will be achieved as planned. Controlling means being sure that progress is being made toward objectives and taking corrective action as necessary. The utility manager is directly involved in controlling the treatment process to ensure that water or wastewater is being properly treated and to make sure that the utility is meeting its short- and long-term goals.

QUESTIONS

Write your answers in a notebook and then compare your answers with those on page 50.

2.0A What are the functions of a utility manager?

2.0B In small communities, what does the community depend on the utility manager to do?

3 PLANNING[1]

A very large portion of any manager's typical work day will be spent on activities that can be described as planning activities since nearly every area of a manager's responsibilities require some type of planning.

Planning is one of the most important functions of utility management and one of the most difficult. Communities must have good, safe, drinking water and the management of water or wastewater utilities must include building the resources and financial capability to provide for future needs. The utility must plan for future growth, including industrial development, and be ready to provide the water and waste treatment systems that will be needed as the community grows. The most difficult problem for some small communities is recognizing and planning for a decline in population. The utility manager must develop reliable information to plan for growth or decline. Decisions must be made about goals, both short- and long-term. The manager must prepare plans for the next two years and the next 10 to 20 years. Remember that utility planning should include operational personnel, local officials (decision makers), and the public. Everyone must understand the importance of planning and be willing to contribute to the process.

Operation and maintenance of a utility also involves planning by the utility manager. A preventive maintenance program should be established to keep the system performing as intended and to protect the community's investment in water or wastewater facilities. (Section 10 describes the various types of maintenance and the benefits of establishing maintenance programs.)

The utility also must have an emergency response plan to deal with natural or human disasters. Without adequate planning your utility will be facing system failures, inability to meet compliance regulations, and inadequate service capacity to meet community needs. Plan today and avoid disaster tomorrow. (Section 11, "Emergency Response Plan," describes the basic elements of an emergency operations plan.)

4 ORGANIZING[2]

A utility should have a written organizational plan and written policies. In some communities the organizational plan and policies are part of the overall community plan. In either case, the utility manager and all plant personnel should have a copy of the organizational plan and written policies of the utility.

The purpose of the organizational plan is to show who reports to whom and to identify the lines of authority. The organizational plan should show each person or job position in the organization with a direct line showing to whom each person reports in the organization. Remember, an employee can serve only one supervisor (unity of command) and each supervisor should ideally manage only six or seven employees. The organizational plan should include a job description for each of the positions on the organizational chart. When the organizational plan is in place, employees know who is their immediate boss and confusion about job tasks is eliminated. A sample organizational plan is shown in Figure 1. The basic job duties for some typical utility positions are described in Table 2.

To understand organization and its role in management, we need to understand some other terms including authority, responsibility, delegation, and accountability. AUTHORITY means the power and resources to do a specific job or to get that job done. Authority may be given to an employee due to their position in the organization (this is formal authority) or authority may be given to the employee informally by their co-workers when the employee has earned their respect. RESPONSIBILITY may be described as answering to those above in the chain of command to explain how and why you have used your authority. DELEGATION is the act in which power is given to another person in the organization to accomplish a specific job. Finally when a manager gives power/responsibility to an employee, then the employee is held ACCOUNTABLE[3] for the results.

Organization and effective delegation are very important to keep any utility operating efficiently. Effective delegation is uncomfortable for many managers since it requires giving up power and responsibility. Many managers believe that they can do the job better than others, they believe that other employees are not well trained or experienced, and they are afraid of mistakes. The utility manager retains some responsibility even after delegating to another employee and, therefore, the manager is often reluctant to delegate or may delegate the responsibility but not the authority to get the job done. For the utility manager, good organization means that employees are ready to accept responsibility and have the power and resources to make sure that the job gets done.

Employees should not be asked to accept responsibilities for job tasks that are beyond their level of authority in the organizational structure. For example, an operator or lead utility worker should not be asked to accept responsibility for additional lab testing. The responsibility for additional lab testing must be delegated to the lab supervisor. Authority and responsibility must be delegated properly to be effective. When these three components—proper job assignments, au-

---

[1] Planning. Management of utilities to build the resources and financial capability to provide for future needs.
[2] Organizing. Deciding who does what work and delegating authority to the appropriate persons. A utility should have a written organizational plan and written policies.
[3] Accountability. When a manager gives power/responsibility to an employee, the employee ensures that the manager is informed of results or events.

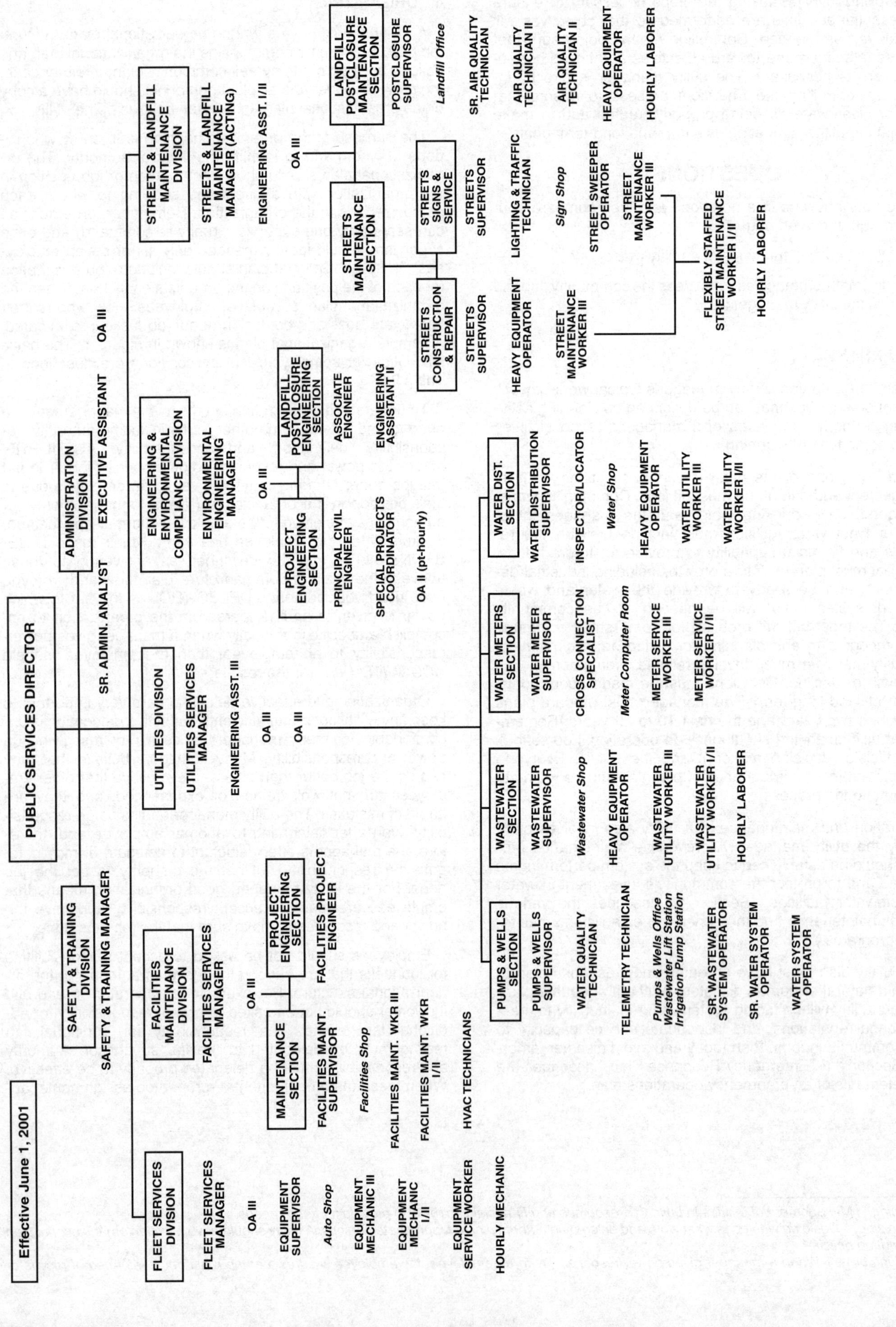

4 Utility Management

Fig. 1 *Organizational chart for medium-sized utility*
(Courtesy of City of Mountain View, California)

TABLE 2   JOB DUTIES FOR STAFF OF A MEDIUM-SIZED UTILITY

Job Title	Job Duties
Superintendent	Responsible for administration, operation, and maintenance of entire facility. Exercises direct authority over all plant functions and personnel.
Assistant Superintendent	Assists Superintendent in review of operation and maintenance function, plans special operation and maintenance tasks.
Clerk/Typist	Performs all clerical duties.
Operations Supervisor	Coordinates activities of plant operators and other personnel. Prepares work schedules, inspects plant, and makes note of operational and maintenance requirements.
Lead Utility Worker	Supervises operations and manages all operators.
Utility Worker II (Journey Level)	Controls treatment processes. Collects samples and delivers them to the lab for analysis. Makes operational decisions.
Utility Worker I	Performs assigned job duties.
Maintenance Supervisor	Supervises all maintenance for plant. Plans and schedules all maintenance work. Responsible for all maintenance records.
Maintenance Foreman	Supervises mechanical maintenance crew. Performs inspections and determines repair methods. Schedules all maintenance including preventive maintenance.
Maintenance Mechanic II	Selects proper tools and assigns specific job tasks. Reports any special considerations to Foreman.
Maintenance Mechanic I	Performs assigned job duties.
Electrician II	Schedules and coordinates electrical maintenance with other planned maintenance. Plans and selects specific work methods.
Electrician I	Performs assigned job duties.
Chemist	Directs all laboratory activities and makes operational recommendations to Operations Supervisor. Reports and maintains all required laboratory records. Oversees laboratory quality control.
Laboratory Technician	Performs laboratory tests. Manages day-to-day laboratory operations.

thority, and responsibility—are all present, the supervisor has successfully delegated. The success of delegation is dependent upon all three components.

An important and often overlooked part of delegation is *follow-up* by the supervisor. A good manager will delegate and follow up on progress to make sure that the employee has the necessary resources to get the job done. Well-organized managers can delegate effectively and do not try to do all the work themselves, but are responsible for getting good results. The Management Muddle No. 1 that follows describes what can happen when delegation is improperly conducted, and illustrates how an organizational plan can prevent disaster.

Management Muddle No. 1

The City Manager of Pleasantville calls the Director of Public Works and asks for a report on the need for and cost of a new sludge truck to be presented at the September 13 meeting of the City Council. The Director of Public Works calls the Plant Manager and asks for a report on the need and cost for a new sludge truck with a deadline of September 12. The Plant Manager calls the Lead Utility Worker, an operator, who has been asking for a new sludge truck and has been looking into the details. The Plant Manager requests that the Lead Utility Worker provide a report on September 12 about the purchase of the sludge truck. The Lead Utility Worker gathers all the notes and hand writes a report identifying the need for the truck, the features required, and the cost. The Lead Utility Worker takes the report to City Hall to be typed and leaves it with a secretary on September 12. On September 13, the City Manager is preparing for the City Council meeting and does not have the report. Who is responsible? Who is accountable? How could this situation have been avoided?

Responsibility: The Lead Utility Worker's responsibility has been carried out with the authority and resources made available. Both the Director of Public Works and the Plant Manager failed to follow up on the report on September 12. No one informed the Lead Utility Worker that the report must be presented to the City Council on September 13, nor was the Lead Utility Worker supplied with the resources for getting the report in final form. However, the City Manager is ultimately responsible for reporting to the City Council.

Accountability: Starting with the Lead Utility Worker and working upward, each employee is accountable to his or her supervisor and should have communicated the status of the report.

How to avoid this situation: Good communication and follow-up by each of these supervisors could have prevented this situation completely. The City Manager should have asked to see the report on September 12; the Director of Public Works should have asked the Plant Manager to deliver the report no later than September 11; and the Plant Manager should have asked the Lead Utility Worker to submit the typed report (to the Plant Manager) no later than September 10. When delegating this task, the Plant Manager should have arranged for a secretary or clerk to assist the Lead Utility Worker in typing the report. Providing clerical support enables the Lead Utility Manager to complete the assigned task in a timely manner.

At each step in this chain of delegation, setting an early deadline gives the individual receiving the report an opportunity to review the document and make revisions, if necessary, before forwarding it up the chain of authority and ensures that the report reaches the City Manager no later than September 12.

QUESTIONS

Write your answers in a notebook and then compare your answers with those on pages 50 and 51.

3.0A Who must be included in utility planning?

4.0A What is the purpose of an organizational plan?

4.0B Why is it sometimes difficult or uncomfortable for supervisors or managers to delegate effectively?

4.0C What is an important and often overlooked part of delegation?

5 STAFFING

5.0 The Utility Manager's Responsibilities

The utility manager is also responsible for staffing, which includes hiring new employees, training employees, and evaluating job performance. The utility should have established procedures for job hiring which include requirements for advertising the position, application procedures, and the procedures for conducting interviews.

In the area of staffing, more than any other area of responsibility, a manager must be extremely cautious and consider the consequences before taking action. Personnel management practices have changed dramatically in the past few years and continue to be redefined almost daily by the courts. A manager who violates an employee's or job applicant's rights can be held both personally and professionally liable in court. Throughout this section on staffing you will repeatedly find references to two terms: **job-related** and **documentation**. These are key concepts in personnel management today. Any personnel action you take must be job-related, from the questions you ask during interviews to disciplinary actions or promotions. And while almost no one wants more paper work, documentation of personnel actions detailing what you did, when you did it, and why you did it (the reasons will be job-related, of course) is absolutely essential. There is no way to predict when you might be called upon to defend your actions in court. Good records not only serve to refresh your memory about past events but can also be used to demonstrate your pattern of lawful behavior over time.

NOTICE

The information provided in this section on staffing should **not** be viewed as **legal advice**. The purpose of this section is simply to identify and describe in general terms the major components of a utility manager's responsibilities in the area of staffing. One issue, harassment, is discussed in somewhat greater detail to illustrate the broad scope of a manager's responsibilities within a single policy area. Personnel administration is affected by many federal and state regulations. Legal requirements of legislation such as the Americans With Disabilities Act (ADA), Equal Employment Opportunity (EEO) Act, Family and Medical Leave Act, and wages and hours laws are complex and beyond the scope of this manual. If your utility does not have established personnel policies and procedures, consider getting help from a labor law attorney to develop appropriate policies. At the very least, you should get help from a recruitment specialist to develop and document hiring procedures that meet the federal guidelines for Equal Employment Opportunity.

5.1 How Many Employees Are Needed?

There is a common tendency for organizations to add personnel in response to changing conditions without first examining how the existing workforce might be reorganized to achieve greater efficiency and meet the new work demands. In water and wastewater utilities, aging of the system, changes in use, and expansion of the system often mean changes in the operation and maintenance tasks being performed. The manager of a utility should periodically review the agency's work requirements and staffing to ensure that the utility is operating as efficiently as possible. A good time to conduct such a review is during the annual budgeting process or when you are considering hiring a new employee because the workload seems to be greater than the current staff can adequately handle.

The staffing analysis procedure outlined in this section illustrates how to conduct a comprehensive analysis of the type needed for a complete reorganization of the agency. In practice, however, a complete reorganization may not be desirable or even possible. Frequent organizational changes can make employees anxious about their jobs and may interfere with their work performance. Some employees show strong resistance to any change in job responsibilities. Nonetheless, by thoroughly examining the functions and staffing of the utility on a periodic basis, the manager may spot trends (such as an increase in the amount of time spent maintaining certain equipment or portions of the system) or discover inefficiencies that could be corrected over a period of time.

The first step in analyzing the utility's staffing needs is to prepare a detailed list of all the tasks to be performed to operate and maintain the utility. Next, estimate the number of staff hours per year required to perform each task. Be sure to include the time required for supervision and training.

When you have completed the task analysis, prepare a list of the utility's current employees. Assign tasks to each employee based on the person's skills and abilities. To the extent possible, try to minimize the number of different work activities assigned to each person but also keep in mind the need to provide opportunities for career advancement. One full-time staff year equals 260 days, including vacation and holiday time: (52 wk/yr)(5 days/wk) = 260 days/yr.

You can expect to find that this "ideal" staffing arrangement does not exactly match up with your current employees' job assignments. Most likely, you will also find that the number of staff hours required does not exactly equal the number of staff hours available. Your responsibility as a manager is to create the best possible fit between the work to be done and the personnel/skills available to do it. In addition to shifting work assignments between employees, other options you might consider are contracting out some types of work, hiring part-time or seasonal staff, or setting up a second shift (to make fuller use of existing equipment). Of course, you may find that it is time to hire another full- or part-time operator.

5.2 Qualifications Profile

Hiring new employees requires careful planning before the personal interview process. In an effort to limit discriminatory hiring practices, the law and administrative policy have carefully defined the hiring methods and guidelines employers may use. The selection method and examination process used to evaluate applicants must be limited to the applicant's knowledge, skills, and abilities to perform relevant job-related activities. In all but rare cases, factors such as age and level of education may not be used to screen candidates in place of performance testing. A description of the duties and qualifications for the job must be clearly defined in writing. The job description may be used to develop a qualifications profile. This qualifications profile clearly and precisely identifies the required job qualifications. All job qualifications must be relevant to the actual job duties that will be performed in that position. The following list of typical job qualifications may be used to help you develop your own qualifications profiles with advice from a recruitment specialist.

8 Utility Management

1. General Requirements:
   a. Knowledge of methods, tools, equipment, and materials used in water/wastewater utilities,
   b. Knowledge of work hazards and applicable safety precautions,
   c. Ability to establish and maintain effective working relations with employees and the general public, and
   d. Possession of a valid state driver's license for the class of equipment the employee is expected to drive.

2. General Educational Development:
   a. Reasoning: Apply common sense understanding to carry out instructions furnished in oral, written, or diagrammatical form,
   b. Mathematical: Use a pocket calculator to make arithmetic calculations relevant to the utility's operation and maintenance processes, and
   c. Language: Communicate with fellow employees and train subordinates in work methods. Fill out maintenance report forms.

3. Specific Vocational Preparation: Three years of experience in water/wastewater utility operation and maintenance.

4. Interests: May or may not be relevant to knowledge, skills, and ability; for example, an interest in activities concerned with objects and machines, ecology, or business management.

5. Temperament: Must adjust to a variety of tasks requiring frequent change and must routinely use established standards and procedures.

6. Physical Demands: Medium to heavy work involving lifting, climbing, kneeling, crouching, crawling, reaching, hearing, and seeing. Must be able to lift and carry _____ number of pounds for a distance of _____ feet.

7. Working Conditions: The work is outdoors and involves wet conditions, cramped and awkward spaces, noise, risks of bodily injury, and exposure to weather.

QUESTIONS

Write your answers in a notebook and then compare your answers with those on page 51.

5.0A What do staffing responsibilities include?

5.0B What are two key personnel management concepts a manager should always keep in mind?

5.1A List the steps involved in a staffing analysis.

5.2A What is a qualifications profile?

5.3 Applications and the Selection Process

5.30 Advertising the Position

To advertise a job opening, first prepare a written description of the required job qualifications, compensation, job duties, selection process, and application procedures (with a closing date). The utility should have established procedures about how to advertise the position and conduct the application process. The application procedure may require that the job be posted first within the utility to allow existing personnel first chance at the job opportunity.

5.31 Paper Screening

The next step in the selection process is known as paper screening. The personnel department and the utility manager review each application and eliminate those who are not qualified. The qualified applicants may be given examinations to verify their qualifications. Usually the top three to twelve applicants are selected for an interview, depending on the agency's preference.

5.32 Interviewing Applicants

The purpose of the job interview is to gain additional information about the applicants so that the most qualified person can be selected. The utility manager should prepare for the interview in advance. Review the background information on each applicant. Draw up a list of job-related questions that will be asked of each applicant. During the interviews, briefly note the answers each applicant gives.

It used to be thought that the best way to learn about applicants was to give them plenty of time to talk about themselves because the content and type of information applicants volunteer might provide a deeper insight into the person and what type of employee they will become. Be very careful about open-ended, unstructured conversations with job applicants, even the friendly remarks you make initially to put the applicant at ease during the interview. If the applicant begins to volunteer information you could not otherwise legally ask for (such as marital status, number of children, religious affiliation, or age), be polite but firm in promptly redirecting the conversation. Even if this information was provided to you voluntarily, an applicant who did not get the job could later allege that you discriminated against them based on age or religion.

The only type of information you may legally request is information about the applicant's job skills, abilities, and experience relating directly to the job for which the person is applying. You must always be sensitive to the civil rights of the applicant and the affirmative action policies of the utility, which is another good reason to prepare a list of questions before the interview process begins. Structure the questions so that you avoid simple yes-and-no answers. Table 3 summarizes acceptable and unacceptable pre-employment inquiries to guide you in developing a good list of questions.

If other utility staff members are participating in the interviews, their participation should be confined to the preselected questions. Under no circumstances should front line employees conduct interviews in the absence of the manager or another person knowledgeable about personnel policies and practices.

The interview should be conducted in a quiet room without interruptions. Most applicants will be nervous, so start the interview on a positive note with introductions and some general remarks to put the applicant at ease. Explain the details of the job, working conditions, wages, benefits, and potential for advancement. Allow the applicant a chance to ask questions about the job. Ask each applicant the questions you have prepared and jot down brief notes on their responses. Taking notes while interviewing is awkward for some people but it becomes easier with practice. Notes are important because after interviewing several candidates you may not be able to remember what each one said. Also, as mentioned earlier, notes taken at the time of an interview can be valuable evidence in court if an unsuccessful applicant files a lawsuit for unfair hiring practices. At the end of the interview, tell the applicant when a decision will be made and how the applicant will be informed of the decision.

If an applicant's responses during the interview indicate that the person clearly is not qualified for this job but may be qualified for another job, briefly describe the other opportunity and how the person can apply for that position. The applicant may ask to be interviewed immediately for the second position. However, do not violate the utility's hiring procedures for the convenience of a job applicant. The same sequence of hiring

TABLE 3   ACCEPTABLE AND UNACCEPTABLE PRE-EMPLOYMENT INQUIRIES[a]

Acceptable Pre-Employment Inquiries	Subject	Unacceptable Pre-Employment Inquiries
"Have you ever worked for this agency under a different name?"	NAME	Former name of applicant whose name has been changed by court order or otherwise.
Applicant's place of residence. How long applicant has been resident of this state or city.	ADDRESS OR DURATION OF RESIDENCE	
"If hired, can you submit a birth certificate or other proof of U.S. citizenship or age?"	BIRTHPLACE	Birthplace of applicant. Birthplace of applicant's parents, spouse, or other relatives. Requirement that applicant submit a birth certificate, naturalization, or baptismal record.
"If hired, can you furnish proof of age?" /or/ Statement that hire is subject to verification that applicant's age meets legal requirements.	AGE	Questions which tend to identify applicants 40 to 64 years of age.
Statement by employer of regular days, hours, or shift to be worked.	RELIGIOUS	Applicant's religious denomination or affiliation, church, parish, pastor, or religious holidays observed. "Do you attend religious services /or/ a house of worship?" Applicant may not be told, "This is a Catholic/Protestant/Jewish/atheist organization."
	RACE OR COLOR	Complexion, color of skin, or other questions directly or indirectly indicating race or color.
Statement that photograph may be required after employment.	PHOTOGRAPH	Requirement that applicant affix a photograph to his/her application form. Request applicant, at his/her option, to submit photograph. Requirement of photograph after interview but before hiring.
Statement by employer that, if hired, applicant may be required to submit proof of eligibility to work in the United States.	CITIZENSHIP	"Are you a United States citizen?" Whether applicant or applicant's parents or spouse are naturalized or native-born U.S. citizens. Date when applicant or parents or spouse acquired U.S. citizenship. Requirement that applicant produce naturalization papers or first papers. Whether applicant's parents or spouse are citizens of the U.S.
Applicant's work experience. Applicant's military experience in armed forces of United States, in a state militia (U.S.), or in a particular branch of the U.S. armed forces.	EXPERIENCE	"Are you currently employed?" Applicant's military experience (general). Type of military discharge.

10 Utility Management

TABLE 3 ACCEPTABLE AND UNACCEPTABLE PRE-EMPLOYMENT INQUIRIES (continued)

Acceptable Pre-Employment Inquiries	Subject	Unacceptable Pre-Employment Inquiries
Applicant's academic, vocational, or professional education; schools attended.	EDUCATION	Date last attended high school.
Language applicant reads, speaks, or writes fluently.	NATIONAL ORIGIN OR ANCESTRY	Applicant's nationality, lineage, ancestry, national origin, descent, or parentage.
		Date of arrival in United States or port of entry; how long a resident.
		Nationality of applicant's parents or spouse; maiden name of applicant's wife or mother.
		Language commonly used by applicant. "What is your mother tongue?"
		How applicant acquired ability to read, write, or speak a foreign language.
	CHARACTER	"Have you ever been arrested?"
Names of applicant's relatives already employed by the agency.	RELATIVES	Marital status or number of dependents.
		Name or address of relative, spouse, or children of adult applicant.
		"With whom do you reside?"
		"Do you live with your parents?"
Organizations, clubs, professional societies, or other associations of which applicant is a member, excluding any names the character of which indicate the race, religious creed, color, national origin, or ancestry of its members.	ORGANIZATIONS	"List all organizations, clubs, societies, and lodges to which you belong."
"By whom were you referred for a position here?"	REFERENCES	Requirement of submission of a religious reference.
"Do you have any physical condition that may limit your ability to perform the job applied for?"	PHYSICAL CONDITION	"Do you have any physical disabilities?"
		Questions on general medical condition.
Statement by employer that offer may be made contingent on passing a physical examination.		Inquiries as to receipt of Workers' Compensation.
Notice to applicant that any misstatements or omissions of material facts in his/her application may be cause for dismissal.	MISCELLANEOUS	Any inquiry that is not job-related or necessary for determining an applicant's eligibility for employment.

[a] Courtesy of Marion B. McCamey, Affirmative Action Officer, California State University, Sacramento, CA.

procedures should be followed each time a position is filled. Tell this applicant that it will still be necessary to apply for the other position and that another interview may or may not follow, depending on the qualifications of the other applicants for the position.

5.33 Selecting the Most Qualified Candidate

Once the interviews are over, the job of evaluating and selecting the successful candidate begins. Review your interview notes and check the candidates' references. Checking references will verify the job experience of the applicant and may provide insight into the applicant's work habits. Questions you might ask previous employers include: Was the employee reliable and punctual? How well did the employee relate to co-workers? Did the employee consistently practice safe work procedures? Would you rehire this employee?

The rights of certain "protected groups" in the workforce today, such as minorities, women, disabled persons, persons over 40 years of age, and union members, are protected by law. A manager's responsibilities regarding protected groups begins with the hiring process and continues for as long as the employer/employee relationship lasts. The best principle to deal with protected groups (and all other employees, for that matter) is the "OUCH" principle. The OUCH principle says that when you hire new employees or delegate job tasks to current employees, you must be OBJECTIVE, UNIFORM in your treatment of applicants or employees, CONSISTENT with utility policies, and HAVE JOB RELATEDNESS. If you don't manage with all of these characteristics, you may find yourself in a "hurting" position with regard to protected workers.

Objectivity is the first hurdle. Often the physical characteristics of a person, such as large or small size, may make a person seem more or less job capable. However, many utility agency jobs are done with power tools or other technology that allows all persons, regardless of size, to manage most tasks. Try to remain objective but reasonable in assessing job applicants and making job assignments.

Uniform treatment of job applicants and employees is necessary to protect yourself and other employees. Nothing will destroy morale more quickly than unequal treatment of employees. Your role as a manager is to consistently apply the policies and procedures that have been adopted by the utility. Often, policies and procedures exist that are not popular and may not even be appropriate. However, the job of the utility manager is to consistently uphold and apply the policies of the utility.

The last part of the OUCH principle is having job relatedness. Any hiring decision must be based on the applicant's qualifications to meet the specific job requirements and any job assignment given to an employee must be related to that employee's job description. Extra assignments, such as buying personal gifts for the boss's family or washing the boss's car, are not appropriate. These types of job assignments will eventually catch up to the manager and can be particularly embarrassing if the public gets involved. So to protect yourself and your utility, remember the OUCH principle as you hire and manage your employees.

Once you have made your selection, the applicant is usually required to pass a medical examination. When this has been successfully completed and the applicant has accepted the position, notify the other applicants that the position has been filled.

5.4 New Employee Orientation

During the first day of work, a new employee should be given all the information available in written and verbal form on the policies and practices of the utility including compensation, benefits, attendance expectations, alcohol and drug testing (if the utility does this), and employer/employee relations. Answer any questions from the new employee at this time and try to explain the overall structure of the utility as well as identify who can answer employee questions when they arise. Introduce the new employee to co-workers and tour the work area. Every utility should have a safety training session for all new employees and specific safety training for some job categories. Provide safety training (see Section 13, "Safety Program") for new employees on the first day of employment or as soon thereafter as possible. Establishing safe work practices is a very important function of management.

QUESTIONS

Write your answers in a notebook and then compare your answers with those on page 51.

5.3A   What is the purpose of a job interview?

5.3B   List four "protected groups."

5.3C   What does the "OUCH" principle stand for?

5.4A   When should a new employee's safety training begin?

5.5 Employment Policies and Procedures

5.50 Probationary Period

Many employers now use a probationary period for all new employees. The probationary period is typically three to six months but may be as long as a year. This period begins on the first day of work. Management may reserve the right to terminate employment of the person with or without cause during this probationary period. The employee must be informed of this probationary period and must understand that successful completion of the probationary period is required in order to move into regular employment status.

The probationary period provides a time during which both the employer and employee can assess the "fit" between the job and the person. Normally a performance evaluation is completed near the end of the probationary period. A satisfactory performance evaluation is the mechanism used to move an employee from probationary status into regular employment.

12 Utility Management

5.51 Compensation

The compensation an employee receives for the work performed includes satisfaction, recognition, security, appropriate pay, and benefits. All are important to keep good employees satisfied. Salaries should be a function of supply and demand. Pay should be high enough to attract and retain qualified employees. Salaries are usually determined by the governing body of the utility in negotiation with employee groups, when appropriate. The salary structure should meet all state and federal regulations and accurately reflect the level of service given by the employee. A survey of salaries from other utilities in the area may provide valuable information in the development of a salary structure.

The benefits supplied by the employer are an important part of the compensation package. Benefits generally include the following: retirement, health insurance, life insurance, employer's portion of social security, holiday and vacation pay, sick leave, personal leave, parental leave, worker's compensation, and protective clothing. Many employers now provide dental and/or vision insurance, long-term disability insurance, educational bonus or costs and leave, bereavement leave, and release time for jury duty. Some employers also include in their benefit package cash bonus programs and longevity pay. The value of an employee's entire benefit package is often computed and printed on the pay stub as a reminder that salary alone is not the only compensation being provided.

5.52 Training and Certification[4]

Training has become an ongoing process in the workplace. The utility manager must provide new employee training as well as ongoing training for all employees. Safety training is particularly important for all utility operators and staff members and is discussed in detail later in Section 12. Certified operators earn their certificates by knowing how to do their jobs safely. Preparing for certification examinations is one means by which operators learn to identify safety hazards and to follow safe procedures at all times under all circumstances.

Although it is extremely important, safety is not the only benefit of a certification program. Other benefits include protection of the public's investment in utility facilities and employee pride and recognition. Vast sums of public funds have been invested in the construction of water and wastewater treatment facilities. Certification of operators assures utilities that these facilities will be operated and maintained by qualified operators who possess a certain level of competence. These operators should have the knowledge and skills not only to prevent unnecessary deterioration and failure of the facilities, but also to improve operation and maintenance techniques.

Achievement of a level of certification is a public acknowledgment of a water supply system or wastewater treatment plant operator's skills and knowledge. Presentation of certificates at an official meeting of the governing body will place the operators in a position to receive recognition for their efforts and may even get press coverage and public opinion that is favorable. An improved public image will give the certified operator more credibility in discussions with property owners.

Recognition for their personal efforts will raise the self-esteem of all certified operators. Certification will also give water and wastewater treatment plant operators an upgraded image that has been too long denied them. If properly publicized, certification ceremonies will give the public a more accurate image of the many dedicated, well-qualified operators working for them. Certification provides a measurable goal that operators can strive for by preparing themselves to do a better job. Passing a certification exam should be recognized by an increase in salary and other employee benefits.

Most states and Canadian provinces now require that water and wastewater plant operators be certified. To maintain current certification, these operators must complete additional training classes every one to five years. In the environmental field, new technologies and regulations require operators to attend training to keep up with their field. The utility manager has the responsibility to provide employees with high-quality training opportunities. Many types of training are available to meet the different training needs of utility operators, for example, in-house training, training conducted by training centers, professional organizations, engineering firms, or regulatory agencies, and correspondence courses such as this one by the Office of Water Programs.

ABC stands for the Association of Boards of Certification for Operating Personnel in Water Utilities and Pollution Control Systems. If you wish to find out how operators can become certified in your state or province, contact:

Executive Director, ABC
208 Fifth Street
Ames, IA 50010-6259
Phone: (515) 232-3623

ABC will provide you with the name and address of the appropriate contact person.

One area of training that is frequently overlooked is training for supervisors. Managing people requires a different set of skills than performing the day-to-day work of operating and maintaining a water or wastewater facility. Supervisors need to know how to communicate effectively and how to motivate others, as well as how to delegate responsibility and hold people accountable for their performance. Supervisors share management's responsibility for fair and equitable treatment of all workers and are required to act in accordance with applicable state and federal personnel regulations. Making the

---

[4] *Certification Examination. An examination administered by a state agency or professional association that managers take to indicate a level of professional competence. Operator certification is mandatory in the United States for the Chief Operators of water treatment plants, water distribution systems, and wastewater treatment plants.*

transition from operator to supervisor also requires a change in attitude. A supervisor is part of the management team and is therefore obliged to promote the best interests of the utility at all times. When the interests of the utility conflict with the desires of one or more employees, the supervisor must support management's decisions and policies regardless of the supervisor's own personal opinion about the issue. It is the responsibility of the utility manager to ensure that supervisors receive appropriate training in all of these areas.

Training on how to motivate people, deal with co-workers, and supervise or manage people working for you has become a very highly specialized field of training. These are complex topics that are beyond the scope of this manual. If you have a need for or wish to learn more about how to deal with people, consider enrolling in courses or reading books on supervision or personnel management.

QUESTIONS

Write your answers in a notebook and then compare your answers with those on page 51.

5.5A What is the purpose of a probationary period for new employees?

5.5B What kind of compensation does an employee receive for work performed?

5.5C Why should utility managers provide training opportunities for employees?

5.53 Performance Evaluation

Most organizations conduct some type of performance evaluation, usually on an annual basis. The evaluation may be written and/or oral; however, a written evaluation is strongly recommended because it will provide a record of the employee's performance. Documentation of this type may be needed in the future to support taking disciplinary action if the employee's performance consistently fails to meet expectations. The evaluation of employee performance can be a challenging task, especially when performance has not been acceptable. However, evaluations are also an opportunity to provide employees with positive feedback and let them know their contributions to the organization have been noticed and appreciated.

A formal performance evaluation typically begins with an employee's immediate supervisor filling out the performance evaluation form (a sample evaluation form is shown in Figure 2). Complete the entire form and be specific about the employee's achievements as well as areas needing improvement. Next, schedule a private meeting with the employee to discuss the evaluation. Give the employee frequent opportunities to be heard and listen carefully. If some of the employee's accomplishments were overlooked, note them on the evaluation form and consider whether this new information changes your overall rating of performance in one or more categories.

After reviewing the employee's performance for the past year, set performance goals for the next year. Be sure to document the goals you have agreed upon. Setting performance goals is particularly important if an employee's performance has been poor and improvement is needed. If appropriate, develop a written performance improvement plan that includes specific dates when you will again review the employee's progress in meeting the performance goals. Some supervisors find it helpful to schedule an informal mid-year meeting with each employee to review their progress and to avoid surprises during the next performance evaluation.

Many employees and managers dread even the thought of a performance evaluation and see it as an ordeal to be endured. When properly conducted, however, a performance review can strengthen the lines of communication and increase trust between the employee and the manager. Use this opportunity to acknowledge the employee's unique contributions and to seek solutions to any problems the employee may be having in completing work assignments. Ask the employee how you can be of assistance in removing any obstacles to getting the job done. If necessary, provide coaching to help the employee understand both how and why certain tasks are performed. Be generous (but sincere) with praise for the good work the employee does well every day and try to keep the employee's shortcomings in perspective. If the person is doing a good job 95 percent of the time, don't let the entire discussion consist of criticism about the remaining 5 percent of the person's job assignments.

At the end of the meeting, ask the employee to sign the evaluation form to acknowledge having seen and discussed it. Give the employee a copy of the evaluation. If the employee disagrees with any part of the evaluation, invite the person to submit a written statement describing the reasons for their disagreement. The written statement should be filed with the completed performance evaluation form.

5.54 Dealing With Disciplinary Problems

Handling employee discipline problems is difficult, even for an experienced manager. But remember, **no discipline problem ever solves itself and the sooner you deal with the problem, the better the outcome will be**. If problem behavior is not corrected, then other employees will become dissatisfied and the problems will increase.

Every utility, no matter how small, should have written employment policies enabling the manager to deal effectively with employee problems. It should also provide a formal complaint or grievance procedure by which employees can have their complaint heard and resolved without fear of retaliation by the supervisor.

Dealing with employee discipline requires tact and skill. You will have to find your own style and then try to stay FLEXIBLE, CALM, and OPEN-MINDED when the situation gets really tough. If you repeatedly find yourself unable to deal successfully with disciplinary problems, consult with the utility's personnel office (if available) or consider enrolling in a management training course designed specifically around strategies and techniques for disciplining employees.

A commonly accepted method for dealing with job-related employee problems is to first discuss the problem with the employee in private. Most employers will give a person two or

EMPLOYEE EVALUATION FORM

Employee Name _____ Date _____
Job Title _____ Department _____

Evaluate the employee on the job now being performed. Circle the number which most nearly expresses your overall judgment. In the space for comments, consider the employee's performance since their last evaluation and make notes about the progress or specific concerns in that area. The care and accuracy of this appraisal will determine its value to you, the employee, and your employer.

JOB KNOWLEDGE: (Consider knowledge of the job gained through experience, education, and special training)

5. Well informed on all phases of work
4. Knowledge thorough enough to perform well without assistance
3. Adequate grasp of essentials, some assistance required
2. Requires considerable assistance to perform
1. Inadequate knowledge

Comments: _____
_____
_____

QUALITY OF WORK: (Consider accuracy and dependability of the results)

5. Exceptionally accurate, practically no mistakes
4. Usually accurate, seldom necessary to check results
3. Acceptable, occasional errors
2. Often unacceptable, frequent errors, needs supervision
1. Unacceptable, too many errors

Comments: _____
_____
_____

INITIATIVE: (Consider the speed with which the employee grasps new job skills)

5. Excellent, grasps new ideas and suggests improvements, is a leader with others
4. Very resourceful, can work unsupervised, manages time well, is reliable
3. Shows initiative on occasion, is reliable
2. Lacks initiative, must be reminded to complete tasks
1. Needs constant prodding to complete job tasks, is unreliable

Comments: _____
_____
_____

Fig. 2 Employee evaluation form

COOPERATION AND RELATIONSHIPS: (Consider manner of handling relationships with co-workers, superiors, and the public)

5. Excellent cooperation and communication with co-workers, supervisors, and others, takes and gives instructions well
4. Gets along well with co-workers
3. Acceptable, usually gets along well, occasionally complains
2. Shows a reluctance to cooperate, complains
1. Very poor cooperation, does not follow instruction, dislikes fellow employees

Comments: _____
_____
_____

ATTENDANCE: (Consider frequency of absences, reasons for absences or tardiness, and promptness in giving notice about absences)

5. Excellent, absent only for emergencies, illness, civic duties, always on time, gives notice when absent
4. Rarely absent or late, always gives notice and good reason
3. Occasionally absent, less important reasons, usually gives notice, but not always in time
2. Often absent, lack of adequate notice or reasons for absenteeism
1. Unexcusable absenteeism, does not give notice, reasons are unacceptable, cannot be depended upon

Comments: _____
_____
_____

OVERALL EVALUATION: Superior _____ Good _____ Satisfactory _____ Unsatisfactory _____

Comments: _____
_____
_____

I hereby certify that this appraisal is my best judgment of the service value of this employee and is based on personal observation and knowledge of the employee's work.

Supervisor's Signature _____ Date _____

I hereby certify that I have personally reviewed this report.

Employee's Signature _____ Date _____

Fig. 2   Employee evaluation form (continued)

three verbal warnings; then the warnings should be written with copies given to the employee. Finally, if the written warnings do not produce positive results, the employee may have to be suspended or dismissed. Your job is to make sure that all warnings are documented with specific descriptions of unsatisfactory behaviors and to make sure that all employees are treated fairly.

Start the disciplinary discussion with a positive comment about the employee. Then identify the problem but keep emotion and blame out of the discussion. The best approach is to state the problem and then ask the employee to suggest a solution. If they respond inappropriately, you must restate the problem and explain that you are trying to find a positive solution that is acceptable to everyone.

Try to keep the discussion focused on solving the problems and do not permit the employee to heap on general complaints, report on what other employees do, or wander from the topic. The following is an example of how you might start the discussion for an employee who is tardy every day. "Joe, you have done a good job in keeping that north side pump station running. You are an asset to this operation. Your tardiness every morning, however, is causing problems. Is there some reason for you to be tardy? We need to find a solution to this problem because your being late creates a bad situation for the night shift. What do you suggest?"

Always remain calm and do not allow yourself to become angry when dealing with an employee about performance issues. If you begin to feel angry and are about to lose control, or if the employee becomes combative or abusive, suggest that the meeting is not producing positive results and schedule an alternative meeting time. Do not let the emotions of the moment carry you into a rage in front of employees. If either you or one of your employees expresses extreme emotions, then the discussion should be postponed until everyone cools down. The following steps may serve as a guide to dealing with confrontation; they apply equally to the employee and the supervisor.

- Maintain an adult approach—positive criticism should be taken/given to improve job skills.
- Create a private environment—job performance issues should be discussed in private between the employee and supervisor.
- Listen very carefully—be sure that both you and the employee understand the situation in the same way. If not, you need to keep talking until both parties are in agreement about the problem and the solution.
- Keep your language appropriate—anger and bad language will cause the situation to escalate. Keep your cool and hold your tongue.
- Stay focused in the present—let go of all the past slights, misunderstandings, and dissatisfaction. Problems must be solved one at a time.
- Aim for a permanent solution—changes in job performance need to be permanent to be effective.

Reports of violence in the workplace appear regularly in newspapers and on television. Managers and supervisors should be alert for signs that an employee might become violent and should take any threat of violence seriously. Common warning signs include abusive language, threatening or confrontational behavior, assault, and brandishing a weapon.

The utility's safe workplace policy should be a "zero tolerance" policy. Any employee who is the target of violent behavior or who witnesses such behavior should be encouraged to immediately report the behavior to a supervisor and the incident should be investigated promptly. If necessary for the immediate safety of other employees, the offending employee should be placed on administrative leave, escorted from the work environment, and permitted to return to work only after the investigation has been completed.

Management Muddle No. 2

Sue has been working for five years as a laboratory technician and was recently passed over for promotion. A lab director who has more college experience than Sue was hired from another plant. Since then Sue's work has not been very good, she has come to work late, and she does not always get all of the lab tests done during her workday. What should be done about Sue? If you were the supervisor, how would you handle this problem with Sue?

Actions: As the supervisor, you should ask Sue to come by your office. In private, you should discuss with Sue why the new lab director was hired. Discuss with her the good work record she has maintained over the past five years, and explain the changes you have seen in her work recently. At this point you might ask her to evaluate her own performance or what she would do if she were in your situation. She might need to express her resentment about the new lab director. If she does, let her ramble and rave just for a few minutes, then stop her. You might say, "OK, you're unhappy, but what are we going to do to change this situation? How can I help you to regain your motivation and improve your work habits?" If possible you might help her figure out a way to continue her college education, go to additional training classes, or reorganize the lab so that her job duties change somewhat. There are many other possibilities for helping Sue to become motivated again but she should be part of the process. It is important to communicate clearly that her job performance must improve.

QUESTIONS

Write your answers in a notebook and then compare your answers with those on page 51.

5.5D Who is the appropriate person to conduct an employee's performance evaluation?

5.5E What should be the attitude of a supervisor or manager when dealing with disciplinary problems?

5.5F What are some common warning signs that an employee could become violent?

5.55 Example Policy: Harassment

Harassment is any behavior that is offensive, annoying, or humiliating to an individual and that interferes with a person's ability to do a job. This behavior is uninvited, often repeated, and creates an uncomfortable or even hostile environment in the workplace. Harassment is not limited to physical behavior but also may be verbal or involve the display of offensive pictures or other images. Sexual harassment is legally defined as unwanted sexual advances, or visual, verbal, or physical conduct of a sexual nature. Any type of harassment is inappropriate in the workplace. A manager's responsibilities with regard to harassment include:

- Establish a written policy (such as the one shown in Figure 3) that clearly defines and prohibits harassment of any type,

- Distribute copies of the harassment policy to all employees and take whatever steps are necessary (small group discussions, general staff meetings, training sessions) to ensure that all employees understand the policy,

- Encourage employees to report incidents of harassment to their immediate supervisor, a manager, or the personnel department,

- Investigate every reported case of harassment, and

- Document all aspects of the complaint investigation, including the procedures followed, statements by witnesses, the complainant, and the accused person, the conclusions reached in the case, and the actions taken (if any).

How do you know when offensive behavior could be considered sexual harassment? Unwelcome sexual advances or other verbal or physical conduct of a sexual nature could be interpreted by an employee as sexual harassment under the following conditions:

- A person is required, or feels they are required, to accept unwelcome sexual conduct in order to get a job or keep a job,

- Decisions about an employee's job or work status are made based upon either the employee's acceptance or rejection of unwelcome sexual conduct, or

- The conduct interferes with the employee's work performance or creates an intimidating, hostile, or offensive working environment.

The following is a list of examples of the kind of behavior that is unacceptable and illegal. It is only a partial list to give you an idea of the scope of the requirements.

- Unwanted hugging, patting, kissing, brushing up against someone's body, or other inappropriate sexual touching,

- Subtle or open pressure for sexual activity,

- Persistent sexually explicit or sexist statements, jokes, or stories,

- Repeated leering or staring at a person's body,

- Suggestive or obscene notes or phone calls, and/or

- Display of sexually explicit pictures or cartoons.

The best way to prevent harassment is to set an example by your own behavior and to keep communication open between employees. In most cases an open discussion with employees about harassment can help everyone understand that innuendo and slurs about a person's race, religion, sex, appearance, or any other personal belief or characteristic are humiliating. The most productive way to control such behavior is by enlisting the help of all employees to feel that they have the right and the responsibility to stop harassment. When you get employees to think about their behavior and how their behavior makes others feel, they will usually realize they should speak up to prevent harassment.

Here is an example of how employees handled a problem of harassment. A group of operators often had coffee in the office of the utility during their morning break. One of the operators often used foul language, which was embarrassing and offensive to one of the operator's co-workers. The manager sent a memo to all employees that mentioned respect for fellow workers and included a reminder about inappropriate language. The next day all of the other operators had taped their copy of the memo to the mailbox of the one operator who was most vocal. These operators found a way to send their message loud and clear—no one wants to work in an environment that is unpleasant to others. As a manager, you must establish an atmosphere that is open, congenial, and harmonious for all employees.

Occasional flirting, innuendo, or jokes may not meet the legal definition of sexual harassment. Nevertheless, they may be offensive or intimidating to others and may be considered harassment. Every employee has a right to a workplace free of discrimination and harassment, and every employee has a responsibility to respect the rights of others.

Managers need to **be aware of** and **take action to prevent** any type of harassment in the workplace. It is not enough to

SUBJECT: **HARASSMENT POLICY AND COMPLAINT PROCEDURE**

NO: _____

PURPOSE:

To establish a strong commitment to prohibit harassment in employment, to define discrimination harassment and to set forth a procedure for investigating and resolving internal complaints of harassment.

POLICY:

Harassment of an applicant or employee by a supervisor, management employee, or co-worker on the basis of race, religion, color, national origin, ancestry, handicap, disability, medical condition, marital status, familial status, sex, sexual orientation, or age will not be tolerated. This policy applies to all terms and conditions of employment, including, but not limited to, hiring, placement, promotion, disciplinary action, layoff, recall, transfer, leave of absence, compensation, and training.

Disciplinary action up to and including termination will be instituted for behavior described in the definition of harassment set forth below:

- Any retaliation against a person for filing a harassment charge or making a harassment complaint is prohibited. Employees found to be retaliating against another employee shall be subject to disciplinary action up to and including termination.

DEFINITION:

Harassment includes, but is not limited to:

A. <u>Verbal Harassment</u>—For example, epithets, derogatory comments or slurs on the basis of race, religious creed, color, national origin, ancestry, handicap, disability, medical condition, marital status, familial status, sex, sexual orientation, or age. This might include inappropriate sex-oriented comments on appearance, including dress or physical features or race-oriented stories.

B. <u>Physical Harassment</u>—For example, assault, impeding or blocking movement, with a physical interference with normal work or movement when directed at an individual on the basis of race, religion, color, national origin, ancestry, handicap, disability, medical condition, marital status, familial status, age, sex, or sexual orientation. This could be conduct in the form of pinching, grabbing, patting, propositioning, leering, or making explicit or implied job threats or promises in return for submission to physical acts.

C. <u>Visual Forms of Harassment</u>—For example, derogatory posters, notices, bulletins, cartoons, or drawings on the basis of race, religious creed, color, national origin, ancestry, handicap, disability, medical conditions, marital status, familial status, sex, sexual orientation, or age.

D. <u>Sexual Favors</u>—Unwelcome sexual advances, requests for sexual favors, and other verbal or physical conduct of a sexual nature which is conditioned upon an employment benefit, unreasonably interferes with an individual's work performance, or creates an offensive work environment.

Fig. 3   Harassment policy
(Courtesy of City of Mountain View, California)

SUBJECT: HARASSMENT POLICY AND COMPLAINT PROCEDURE (continued)

COMPLAINT PROCEDURE:

A. <u>Filing</u>:

An employee who believes he or she has been harassed may make a complaint orally or in writing with any of the following:

1. Immediate supervisor.
2. Any supervisor or manager within or outside of the department.
3. Department head.
4. Employee Services Director (or his/her designee).

Any supervisor or department head who receives a harassment complaint should notify the Employee Services Director immediately.

B. Upon notification of the harassment complaint, the Employee Services Director shall:

1. Authorize the investigation of the complaint and supervise and/or investigate the complaint. The investigation will include interviews with:

    (a) The complainant;
    (b) The accused harasser; and
    (c) Any other persons the Employee Services Director has reasons to believe has relevant knowledge concerning the complaint. This may include victims of similar conduct.

2. Review factual information gathered through the investigation to determine whether the alleged conduct constitutes harassment; giving consideration to all factual information, the totality of the circumstances, including the nature of the verbal, physical, visual, or sexual conduct and the context in which the alleged incidents occurred.

3. Report the results of the investigation and the determination as to whether harassment occurred to appropriate persons, including to the complainant, the alleged harasser, the supervisor, and the department head. If discipline is imposed, the discipline may or may not be communicated to the complainant.

4. If harassment occurred, take and/or recommend to the appropriate department head or other appropriate authority prompt and effective remedial action against the harasser. The action will be commensurate with the severity of the offense.

5. Take reasonable steps to protect the victim and other potential victims from further harassment.

6. Take reasonable steps to protect the victim from any retaliation as a result of communicating the complaint.

DISSEMINATION OF POLICY:

All employees, supervisors, and managers shall be sent copies of this policy.

Effective Date: May, 2000
Revision Date: March 1, 2000

City Manager

Fig. 3 Harassment policy (continued)
(Courtesy of City of Mountain View, California)

simply distribute copies of the utility's harassment policy. If legal action is taken against the utility due to harassment or the existence of a hostile work environment, the utility manager may face both personal and professional liability if it can be shown that the manager *should have known* harassment was occurring or that the manager permitted a hostile environment to continue to exist.

Management Muddle No. 3

The maintenance crew is a group of five men who have been with the utility for many years and are well respected for their work habits. However, in the maintenance shed the walls are covered by calendars of scantily clad women and the language used out in the shed is sometimes pretty rough. One of your operators is a woman; she is well respected by her co-workers and is a very good operator. She comes to your office to complain about the situation in the maintenance shed and demands that you remove the pictures from the walls in the shed. She goes on to report that when she went out to the shed and requested assistance to check on a pump, which was noisy and running hot, she was told "Kiss my_____, toots!" As the manager, what should you do? Should you immediately go to the maintenance shed and rip down the pictures? What should you say to the female operator in your office?

Actions: Your first response should be to reassure your operator that you understand her anger and frustration. Let her know that you will investigate the matter immediately and take action to correct any problems you find. Ask her to write out a complete statement of the facts, including her concerns about the pump, when and how her request for assistance was made, to whom the request was made, who made the offensive remark, the names of any witnesses, and what responses she has gotten from the maintenance crew in the past. Try to establish if this is a one-time response or if this problem has been going on for some time.

Begin your investigation with a trip to the maintenance shed to observe and evaluate what is hanging on the walls. Discuss the situation with the crew. Try to make this an open discussion so that everybody understands how the pictures affect the atmosphere and the image of professionalism of the utility. The best solution is to get the crew to understand how this type of behavior looks to persons outside their own small group and then let them take down the pictures. (Be sure to follow up later to confirm that the calendars or other offensive material has not reappeared.)

Next, set up private interviews with the person accused of making the offensive remark and each of the witnesses. Ask each person to describe the encounter in the maintenance shed and make detailed notes of their responses. (Depending on the complexity of the situation, it may sometimes be appropriate to have each person involved submit a written statement describing what occurred.)

After you have thoroughly investigated the incident, discuss with the crew the use of acceptable language in response to other employees. An open discussion and increased awareness of sexual harassment should be all that is necessary to change this situation. If it is not, arrange for a training program in sexual harassment awareness, if one is available. Be sure to establish a policy on the consequences of inappropriate behavior and be prepared to enforce the policy when needed.

Retaliation against an employee for filing a complaint about harassment or a hostile work environment is also illegal. Some examples of retaliation are demotion, suspension, failure to hire or consider for hire, failure to make impartial employment recommendations or decisions, adversely changing working conditions, spreading rumors, or denying any employment benefit. Retaliation could be the basis for a lawsuit involving not only the person who is accused of retaliation, but also the immediate supervisor, the manager, and the utility.

Most areas of personnel management, including harassment and retaliation issues, are complex and have significant legal consequences for everyone involved. The discussion in this section is *NOT* a complete explanation of harassment or retaliation. If your utility is large enough to employ a personnel specialist or a labor law attorney, ask them to review your staffing policies and procedures and consult with them whenever you have questions about personnel matters. If you manage a small utility and have no in-house sources for technical or legal advice, enroll in appropriate training courses or consider working with an attorney on a contract basis.

5.56 Laws Governing Employer/Employee Relations

Many employers take pride in advertising that they are an "equal opportunity employer," and one often sees this claim in newspaper help wanted ads and other forms of job postings. It means an employer's staffing policies and procedures do not discriminate against anyone based on race, religion, national origin, color, citizenship, marital status, gender, age, Vietnam era or disabled veteran status, or the presence of a physical, mental, or sensory disability. An employer must meet specific requirements of the federal Equal Employment Opportunity Act to be eligible to advertise as an equal opportunity employer. These requirements include adoption of nondiscriminatory personnel policies and procedures and periodic submission of reports of personnel actions for review by the Equal Employment Opportunity Commission.

The Family and Medical Leave Act of 1993 (FMLA) is a federal law that requires all public agencies as well as companies with 50 or more employees to permit eligible employees to take up to 12 weeks of time off in a 12-month period for the following purposes: (1) the employee's own serious health condition, (2) to care for a child following birth or placement for adoption or foster care, or (3) to care for the employee's spouse, child, or parent with a serious health condition. To be eligible to receive this benefit, an employee must have been employed for at least one year prior to the leave. The employer is not required to pay the employee's salary during the time off work, but many employers permit (or require) employees to use accrued sick leave and vacation time during the period of unpaid FMLA leave.

The Americans With Disabilities Act of 1990 (ADA) prohibits employment discrimination based on a person's mental or physical disability. The law applies to employers engaged in an industry affecting commerce who have 15 or more employees.

In general, the ADA defines disability as a physical or mental impairment that substantially limits one or more of the major life activities of an individual. The exact meaning of this definition is evolving as the courts settle lawsuits in which individuals allege they were discriminated against because of a physical or mental disability. The original ADA legislation listed more than 40 specific types of impairments and the courts continue to expand the list.

Under the ADA, employers must make reasonable accommodations to enable a disabled person to function successfully in the work environment, for example, installing a ramp to make facilities accessible to someone in a wheelchair, or restructuring an individual's job or work schedule, or providing an interpreter. The requirements of each situation are unique. In each case, the nature and extent of the disability and the "reasonableness" (including cost factors) of the requested accommodation by the employer must be weighed. Employers are not automatically required to do everything possible to accommodate disabled persons, but rather to take whatever reasonable steps they can to do so.

All of these personnel laws are very complex and managers of any utility that may be covered by them are strongly urged to seek the assistance of an experienced labor law attorney or personnel specialist.

5.57 Personnel Records

A personnel file should be maintained for each utility employee. This file should contain all documents related to the employee's hiring, performance reviews, promotions, disciplinary actions, and any other records of employment-related matters. Since these records often contain sensitive, confidential information, access to personnel records should be closely controlled. (Also see Section 14, "Recordkeeping," for more information about what records should be kept and for how long.)

QUESTIONS

Write your answers in a notebook and then compare your answers with those on page 51.

5.5G What is harassment?

5.5H List three types of behavior that could be considered sexual harassment.

5.5I What is the best way to prevent harassment?

5.5J What is the meaning of "disability" under the Americans With Disabilities Act?

5.6 Unions

Whether your utility operators belong to a union now or may join one in the future, a good employee-management relationship is crucial to keeping an agency functioning properly. Managers, supervisors, crew leaders, and operators all have to work together to develop this relationship.

Most of a manager's union contacts are with a shop steward. The shop steward is elected by the union employees and is their official representative to management and the local union. The steward is in an awkward position because the steward is an employee who is expected to do a full-time job like other employees, while also representing all of the employees. The steward must create an effective link between the utility manager and/or supervisors and the employees.

During contract negotiations between management and the employees' union representatives, management should be in constant consultation with the supervisors. Many employee demands regarding working conditions originate from the supervisor's daily dealings with the employees. An effective supervisor can minimize unreasonable demands. Also any demands that are agreed upon must be implemented and carried out by a supervisor. A supervisor can help both sides reach an acceptable contractual agreement.

Once a contract has been agreed upon by both the union and the utility, the utility manager and the other supervisors must manage the organization within the framework of the contract. Do not attempt to ignore or "get around" the contract even if you disagree with some aspects of it. If you do not understand certain contract provisions, ask for clarification before you begin implementation of those provisions.

Contracts do not change the supervisor's delegated authority or responsibility. Operators must carry out the supervisor's orders and get the work done properly, safely, and within a reasonable amount of time. As a supervisor, you have the right and even the duty to make decisions. However, a contract gives a union the right to protest or challenge your decision. When an operator requires discipline, disciplinary action is a management responsibility.

Handling employee grievances within the framework of a union contract can be a very time-consuming job for the supervisor and the steward. Union contracts usually spell out in great detail the steps and procedures the steward and the supervisor must follow to settle differences. Grievances can develop over disciplinary action, distribution of overtime, transfers, promotions, demotions, and interpretation of labor contracts. The shop steward must communicate complaints and grievances from operators to the supervisor. Then the supervisor and the steward must work together to settle complaints and adjust grievances. When a shop steward and supervisor can work together, the steward can help the supervisor to be an effective manager.

22 Utility Management

An effective manager lets everyone know they are available to discuss problems. **Dealing with grievances as quickly as possible** often prevents small problems from growing into large problems. When a shop steward presents a grievance to you, listen carefully and sympathetically to the steward. Discuss the problem with the employee directly with the help of the shop steward. Try to identify the facts and cause of the problem and keep a written record of your findings. Focus on the problem and do not get caught up in irrelevant issues. Make every effort to settle the grievance quickly and to everyone's satisfaction.

The consequences of any solution to a grievance must be considered and solutions must be consistent and fair to other operators. The solution or settlement should be clear and understandable to everyone involved. Once a solution has been agreed upon, prepare a written summary of the agreement. Review this final report with the shop steward to be sure the intent of the solution is understood and properly documented. The entire grievance procedure must be documented and properly filed, from initial presentation to final solution and settlement.

Union activities are governed by the National Labor Relations Act. When a union attempts to organize the utility's employees, your rights and actions as a manager are also governed by this Act. Be sure to seek competent legal assistance if your experience in dealing with a union is limited or if you have no such experience.

QUESTIONS

Write your answers in a notebook and then compare your answers with those on page 51.

5.6A What is the role of a shop steward?

5.6B How does a union contract affect a supervisor's authority?

END OF LESSON 1 OF 2 LESSONS
ON
UTILITY MANAGEMENT

Please answer the discussion and review questions next.

UTILITY MANAGEMENT
DISCUSSION AND REVIEW QUESTIONS
(Lesson 1 of 2 Lessons)

At the end of each lesson in this manual you will find some discussion and review questions. The purpose of these questions is to indicate to you how well you understand the material in the lesson. Write the answers to these questions in your notebook before continuing.

1. What are the different types of demands on a utility manager?
2. List the basic functions of a manager.
3. What can happen without adequate utility planning?
4. What is the purpose of an organizational plan?
5. Define the following terms:
   1. Authority,
   2. Responsibility,
   3. Delegation, and
   4. Accountability.
6. When has a supervisor successfully delegated?
7. What two concepts should a manager keep in mind to avoid violating the rights of an employee or job applicant?
8. Why should a manager thoroughly examine the functions and staffing of the utility on a periodic basis?
9. When hiring new employees, the selection method and examination process used to evaluate applicants must be based on what criteria?
10. Why should you make notes of applicants' responses during job interviews?
11. What information should be provided to a new employee during orientation?
12. What type of training should be provided for supervisors?
13. Why is it important to formally document each employee's performance on a regular basis?
14. How should discipline problems be solved?
15. What steps can you take to help reach a successful resolution to a confrontation with an employee?
16. What is a manager's responsibility for preventing harassment in the workplace?
17. How are employee grievances usually handled under a union contract?

UTILITY MANAGEMENT

(Lesson 2 of 2 Lessons)

6 COMMUNICATION

Good communication is an essential part of good management skills. Both written and oral communication skills are needed to effectively organize and direct the operation of a treatment facility. Remember that communication is a two-part process; information must be given and it must be understood. Good listening skills are as important in communication as the information you need to communicate. As the manager of a water or wastewater utility, you will need to communicate with employees, with your governing body, and with the public. Your communication style will be slightly different with each of these groups but you should be able to adjust easily to your audience.

6.0 Oral Communication

Oral communication may be informal, such as talking with employees, or it may be formal, such as giving a technical presentation. In both cases your words should be appropriate to the audience, for example, avoid technical jargon when talking with nontechnical audiences. As you talk, you should be observing your audience to be sure that what you are saying is getting across. If you are talking with an employee, it is a good idea to ask for feedback from the employee, especially if you are giving instructions. When the employee is talking, watch and listen carefully and clarify areas that seem unclear. Likewise, in a more formal presentation, watching your audience will give you feedback about how well your message is being received. Some tips for preparing a formal speech are given in Table 4.

6.1 Written Communication

Written communication is more demanding than oral communication and requires more careful preparation. Again, keep your audience in mind and use language that will be understood. Written communication requires more organization since you cannot clarify and explain ideas in response to your audience. Before you begin you should have a clear idea of exactly what you wish to communicate, then keep your language as concise as possible. Extra words and phrases tend to confuse and clutter your message. Good writing skills develop slowly, but you should be able to find good writing classes in your community if you need help improving your skills. In addition, many publications and computer software programs are available to assist you in writing memos, letters, press releases, resumes, monitoring reports, monthly reports, and the annual report.

Before you can write a report you must first organize your thoughts. Ask yourself, what is the objective of this report? Am I trying to persuade someone of something? What information is important to communicate in this report? For whom is the report being written? How can I make it interesting? What does the reader want to learn from this report?

After you have answered the above questions, the next step is to prepare a general outline of how you intend to proceed with the preparation of the report. List not only key topics, but try to list all of the related topics. Then arrange the key topics in sequence so there is a workable, smooth flow from one topic to the next. Do not attempt to make your outline perfect. It is just a guide. It should be flexible. As you write you will find that you need to remove nonessential points and expand on more important points.

You might, for example, outline the following points in preparation for writing a report on a polymer testing program.

- A problem condition of high turbidity was discovered
- Polymer testing offered the best means of reducing turbidity
- Funds, equipment, and material were acquired
- Operators were trained
- Tests were conducted
- Results were evaluated and conclusions were reached
- Corrective actions were planned and taken
- Conclusion, the tests did or did not produce the anticipated results or correct the problem

Once you are fairly sure you have included all the major topics you will want to discuss, go through the outline and write down facts you want to include on each topic. As you work through it, you may decide to move material from one topic to another. The outline will help you organize your ideas and facts.

TABLE 4 TIPS FOR GIVING AN ORAL PRESENTATION

1. Arrive early. Give yourself plenty of time to become familiar with the room, practice using your audiovisuals, and make any necessary changes in room setup.
2. Be ready for mistakes. Number the pages in your presentation and your audiovisuals. Check the order of the pages before the meeting begins.
3. Pace yourself. Don't speak too quickly; speak slowly and carefully. Keep a careful eye on audience reaction to be sure that you are speaking at a pace that can be understood.
4. Project yourself. Speak loudly and look at the audience. Do not talk with your back to the audience. Check that those in the back can hear you.
5. Be natural. Try not to read your presentation. Practice ahead of time so that you can speak normally and keep eye contact with your audience.
6. Connect with the audience. Try to smile and make eye contact with audience.
7. Involve the audience. Allow for audience questions and invite their comments.
8. Repeat audience questions. Always repeat the question so everyone can hear and to be sure that you hear the question correctly.
9. Know when to stop. Keep your remarks within the time allocated for your presentation and be aware that long, rambling speeches create a negative impression on the audience.
10. Use readable audiovisuals. Audiovisuals should enhance and reinforce your words. Be sure that all members of the audience can see and read your audiovisuals. Normal typewritten text is not readable on overheads; use large type so everyone can see. Use no more than five to seven key ideas per overhead.
11. Organize your presentation. Prepare an introduction, body of the speech, and conclusion. "Tell them what you're going to say, say it, and then tell them what you said." The presentation should have three to five main points presented in some logical order, for example, chronologically or from simple to complex.

When your outline is complete, you will have the essentials of your report. Now you need to tailor it to the audience that will be reading it. Take a few minutes to think about your audience. What information do they want? What aspects of the topics will they be most interested in reading? Each of the following groups may be interested in specific topics in the report. Consider these interests as you write.

1. Management

   Management will have specific interests that relate to the cost effectiveness of the program. A report to management should include a summary that presents the essential information, procedures used, an analysis of the data (including trends), and conclusions. Be sure to include complete cost information. Did the benefits warrant the costs? As a result of the tests, can future expenses be reduced? Backup information and field data can be included in an appendix for those who want more information.

2. Other Utilities

   Other utilities will be interested in costs but will also want more detailed information about how the program was performed. They will also be interested in the results and benefits of the program. Explain how the tests were done, the procedures, size of the crew, equipment used, source and availability of materials, difficulties encountered and how they were overcome.

3. Citizen Groups

   Citizens' interest will be more general. What is a polymer test and why is it needed? Is the polymer harmless? Will it injure fish or birds? How does the polymer test work? Who pays for the test? How much will it cost to implement results and will they be effective?

Your report may be written to include all of these groups. Adjust the outline to include the topics of interest to each group and identify the topics so readers can find the information most interesting to them. Keep the following information in mind as you write your report:

- Drafts. Good reports are not perfect the first time. Re-read and improve your report several times.
- Facts. Confine your writing to the facts and events that occurred. Include figures and statistics only when they make the report more effective. Include only the relevant facts. Large amounts of data should be put in the appendix. Do not clutter the report with unimportant data.
- Continuity. To be interesting and understood, a report must have continuity. It must make sense to the reader and be organized logically. In the report on the polymer testing, the report should be organized to show you had a problem, you had to find a way to identify where the problem existed, you did the testing, and you identified the problems and the corrective actions.

- Effective. To be effective, a report should achieve the objective for which it was written. In this example we wanted to justify the costs for the program to management, help other utilities in conducting a similar program, and help citizens to understand what we were doing and why.
- Candid. A good report should be frank and straightforward. Keep the language appropriate for the audience. Do not try to impress your readers with technical terms they do not understand. Your purpose is to communicate information. Keep the information accurate and easy to understand.

The annual report is an important part of the management of the utility. It is one of the most involved writing projects that the utility must put together. The annual report should be a review of what and how the utility operated during the past year and it should also include the goals for the next year. In many small communities the annual report may be presented orally to the city council rather than written. If this report is well written, it can be used to highlight accomplishments and provide support for future planning.

The first step to organizing the report is to make a list of three or four major accomplishments of the last year, then make a list of the top three goals for next year. These accomplishments and goals should be the focus of the report. The annual report should be a summary of the expenses, treatment services provided, and revenues generated over the last year. As you organize this information, keep those accomplishments in mind and let the data tell the story of how the utility accomplished last year's tasks. The data by itself may seem boring but as you organize the data it becomes a meaningful description of the year's accomplishments. Conclude with projections for next year. The facts and figures should tell the audience how you plan to accomplish your goals for the next year. The annual report may be simple or complex depending on your community needs. A sample Table of Contents for a medium-sized utility is given in Table 5. When you are finished, the annual report will be a valuable planning tool for the utility and can be used to build support for new projects.

TABLE 5 EXAMPLE OF TABLE OF CONTENTS FOR THE ANNUAL REPORT OF A UTILITY

TABLE OF CONTENTS

Executive Summary

Summary of the Treatment Process Including Flows and Costs

Review of Goals and Objectives for the Year

Special Projects Completed

Professional Awards or Recognition for the Utility or Its Staff

General Operating Conditions Including Regulatory Requirements

Expectations for the Next Year—Goals and Objectives

Recommended Changes for the Utility in Organizing, Staffing, Equipment, or Resources Summary

Appendixes: Operating Data
            Budget
            Information on Special Projects

QUESTIONS

Write your answers in a notebook and then compare your answers with those on page 51.

6.0A What kinds of communication skills are needed by a manager?

6.0B What are the most common written documents that a utility manager must write?

6.0C What should be included in the annual report?

7 CONDUCTING MEETINGS

As a utility manager you will be asked to conduct meetings. These meetings may be with employees, your governing board, the public, or with other professionals in your field. Many new managers fail to prepare for these meetings and the meetings end up as a terrible waste of time. As a manager you need to learn to conduct meetings in a way that is productive and guides the participants into an active role. The following steps should be taken to conduct a productive meeting.

Before the meeting:

- Prepare an agenda and distribute it to all participants.
- Find an adequate meeting room.
- Set a beginning and ending time for the meeting.

During the meeting:

- Start the meeting on time.
- Clearly state the purpose and objectives of the meeting.
- Involve all the participants.
- Do not let one or two individuals dominate the meeting.
- Keep the discussion on track and on time with the agenda.
- When the group makes a decision or reaches consensus, restate your understanding of the results.
- Make clear assignments for participants and review them with everyone during the meeting.

After the meeting:

- Send out minutes of the meeting.
- Send out reminders, when appropriate, about any assignments made for participants, and the next meeting time.

QUESTIONS

Write your answers in a notebook and then compare your answers with those on page 51.

7.0A   With whom may a utility manager be asked to conduct meetings?

7.0B   What should be done before a meeting?

8 PUBLIC RELATIONS

8.0 Establish Objectives

The first step in organizing an effective public relations campaign is to establish objectives. The only way to know whether your program is a success is to have a clear idea of what you expect to achieve—for example, better customer relations, greater water conservation, and enhanced organizational credibility. Each objective must be specific, achievable, and measurable. It is also important to know your audience and tailor various elements of your public relations effort to specific groups you wish to reach, such as community leaders, school children, or the average customer. Your objective may be the same in each case, but what you say and how you say it will depend upon your target audience.

8.1 Utility Operations

Good public relations begin at home. Dedicated, service-oriented employees provide for better public relations than paid advertising or complicated public relations campaigns. For most people, contact with an agency employee establishes their first impression of the competence of the organization, and those initial opinions are difficult to change.

In addition to ensuring that employees are adequately trained to do their jobs and knowledgeable about the utility's operations, management has the responsibility to keep employees informed about the organization's plans, practices, and goals. Newsletters, bulletin boards, and regular, open communication between supervisors and subordinates will help build understanding and contribute to a team spirit.

Despite the old adage to the contrary, the customer is not always right. Management should try to instill among its employees the attitude that while the customer may be confused or unclear about the situation, everyone is entitled to courteous treatment and a factual explanation. Whenever possible, employees should phrase responses as positively, or neutrally, as possible, avoiding negative language. For example, "Your complaint" is better stated as "Your question." "You should have..." is likely to make the customer defensive, while "Will you please..." is courteous and respectful. "You made a mistake" emphasizes the negative, "What we'll do..." is a positive, problem-solving approach.

8.2 The Mass Media

We live in the age of communications, and one of the most effective and least expensive ways to reach people is through the mass media—radio, television, newspapers, and the Internet. Each medium has different needs and deadlines, and obtaining coverage for your issue or event is easier if you are aware of these constraints. Television must have strong visuals, for example. When scheduling a press conference, provide an interesting setting and be prepared to suggest good shots to the reporter. Radio's main advantage over television and newspapers is immediacy, so have a spokesperson available and prepared to give the interview over the telephone if necessary. Newspapers give more thorough, in-depth coverage to stories than do the broadcast media, so be prepared to spend extra time with print reporters and provide written backup information and additional contacts.

It is not difficult to get press coverage for your event or press conference if a few simple guidelines are followed:

1. Demonstrate that your story is newsworthy, that it involves something unusual or interesting.

2. Make sure your story will fit the targeted format (television, radio, newspaper, or the Internet).

3. Provide a spokesperson who is interesting, articulate, and well prepared.

8.3 Being Interviewed

Whether you are preparing for a scheduled interview or are simply contacted by the press on a breaking news story, here are some key hints to keep in mind when being interviewed.

1. Speak in personal terms, free of institutional jargon.

2. Do not argue or show anger if the reporter appears to be rude or overly aggressive.

3. If you don't know an answer, say so and offer to find out. Don't bluff.

4. If you say you will call back by a certain time, do so. Reporters face tight deadlines.

5. State your key points early in the interview, concisely and clearly. If the reporter wants more information, he or she will ask for it.
6. If a question contains language or concepts with which you disagree, don't repeat them, even to deny them.
7. Know your facts.
8. Never ask to see a story before it is printed or broadcast. Doing so indicates that you doubt the reporter's ability and professionalism.

8.4 Public Speaking

Direct contact with people in your community is another effective tool in promoting your utility. Though the audiences tend to be small, a personal, face-to-face presentation generally leaves a strong and long-lasting impact on the listener.

Depending upon the size of the organization, your utility may wish to establish a speaker's bureau and send a list of topics to service clubs in the area. Visits to high schools and college campuses can also be beneficial, and educators are often looking for new and interesting topics to supplement their curriculum.

Effective public speaking takes practice. It is important to be well prepared while retaining a personal, informal style. Find out how long your talk is expected to be, and don't exceed that time frame. Have a definite beginning, middle, and end to your presentation. Visual aids such as charts, slides, or models can assist in conveying your message. The use of humor and anecdotes can help to warm up the audience and build rapport between the speaker and the listener. Just be sure the humor is natural, not forced, and that the point of your story is accessible to the particular audience. Try to keep in mind that audiences only expect you to do your best. They are interested in learning about their utility and will appreciate that you are making a sincere effort to inform them about an important subject.

8.5 Telephone Contacts

First impressions are extremely important, and frequently a person's first contact with your utility is over the telephone. A person who answers the phone in a courteous, pleasant, and helpful manner goes a long way toward establishing a friendly, cooperative atmosphere. Be sure anyone answering telephone inquiries receives appropriate training and conveys a positive image for the utility.

Following a few simple guidelines will help to start your utility off on the right note with your customers:

1. ANSWER CALLS PROMPTLY. Your conversation will get off to a better start if the phone is answered by the third or fourth ring.
2. IDENTIFY YOURSELF. This adds a personal note and lets the caller know whom he or she is talking to.
3. PAY ATTENTION. Don't conduct side conversations. Minimize distractions so you can give the caller your full attention, avoiding repetitions of names, addresses, and other pertinent facts.
4. MINIMIZE TRANSFERS. Nobody likes to get the run-around. Few things are more frustrating to a caller than being transferred from office to office, repeating the situation, problem, or concern over and over again. Transfer only those calls that must be transferred, and make certain you are referring the caller to the right person. Then, explain why you are transferring the call. This lets the caller know you are referring him or her to a co-worker for a reason and reassures the customer that the problem or question will be dealt with. In some cases, it may be better to take a message and have someone return the call than to keep transferring the customer's call.

8.6 Customer Inquiries

No single set of rules can possibly apply to all types of customer questions or complaints about utility service. There are, however, basic principles to follow in responding to inquiries and concerns.

1. BE PREPARED. Your employees should be familiar enough with your utility's organization, services, and policies to either respond to the question or complaint or locate the person who can.
2. LISTEN. Ask the customer to describe the problem and listen carefully to the explanation. Take written notes of the facts and addresses.
3. DON'T ARGUE. Callers often express a great deal of pent-up frustration in their contacts with a utility. Give the caller your full attention. Once you've heard them out, most people will calm down and state their problems in more reasonable terms.
4. AVOID JARGON. The average customer lacks the technical knowledge to understand the complexities of water quality or wastewater treatment. Use plain, nontechnical language and avoid telling the consumer more than he or she needs to know.
5. SUMMARIZE THE PROBLEM. Repeat your understanding of the situation back to the caller. This will assure the customer that you understand the problem and offer the opportunity to clear up any confusion or missed communication.
6. PROMISE SPECIFIC ACTION. Make an effort to give the customer an immediate, clear, and accurate answer to the problem. Be as specific as possible without promising something you can't deliver.

In some cases, you may wish to have a representative of the utility visit the customer and observe the problem first hand. If the complaint involves water quality, take samples if necessary and report back to the customer to be sure the problem has been resolved.

Complaints can be a valuable asset in determining consumer acceptance and pinpointing problems. Customer calls are frequently your first indication that something may be wrong. Responding to complaints and inquiries promptly can save the utility money and staff resources, and minimize the number of customers who are inconvenienced. Still, education can greatly reduce complaints about utility services. Information brochures, utility bill inserts, and other educational tools help to inform customers and avoid future complaints.

8.7 Plant Tours

Tours of water and wastewater treatment plants can be an excellent way to inform the public about your utility's efforts. Political leaders, such as the City Council and members of the Board of Supervisors, should be invited and encouraged to tour the facilities, as should school groups and service clubs.

A brochure describing your utility's goals, accomplishments, operations, and processes can be a good supplement to the tour and should be handed out at the end of the visit. The more visually interesting the brochure is, the more likely that it will be read, and the use of color, photographs, graphics, or other design features is encouraged. If you have access to the necessary equipment, production of a videotape program about the utility can also add interest to the facility tour.

The tour itself should be conducted by an employee who is very familiar with plant operations and can answer the types of questions that are likely to arise. Consider including:

1. A description of the sources of water supply (if appropriate),
2. History of the plant, the years of operation, modifications, and innovations over the years,
3. Major plant design features, including plant capacity and safety features,
4. Observation of the treatment processes,
5. A visit to the laboratory, and
6. Anticipated improvements, expansions, and long-range plans for meeting future service needs.

Plant tours can contribute to a utility's overall program to gain financing for capital improvements. If the City Council or other governing board has seen the treatment process first hand, it is more likely to understand the need for enhancement and support future funding.

QUESTIONS

Write your answers in a notebook and then compare your answers with those on page 52.

8.0A What is the first step in organizing a public relations campaign?

8.1A How can employees be kept informed of the utility's plans, practices, and goals?

8.2A Which news medium is more likely to give a story thorough, in-depth coverage?

8.4A What is the key to effective public speaking?

8.6A How do customer complaints help a utility?

9 FINANCIAL MANAGEMENT

Financial management for a utility should include providing financial stability for the utility, careful budgeting, and providing capital improvement funds for future utility expansion. These three areas must be examined on a routine basis to ensure the continued operation of the utility. They may be formally reviewed on an annual basis or more frequently when the utility is changing rapidly. The utility manager should understand what is required for each of the three areas and be able to develop record systems that keep the utility on track and financially prepared for the future.

9.0 Financial Stability

How do you measure financial stability for a utility? Two very simple calculations can be used to help you determine how healthy and stable the finances are for the utility. These two calculations are the OPERATING RATIO and the COVERAGE RATIO. The operating ratio is a measure of the total revenues divided by the total operating expenses. The coverage ratio is a measure of the ability of the utility to pay the principal and interest on loans and bonds (this is known as DEBT SERVICE[5]) in addition to any unexpected expenses. A utility that is in good financial shape will have an operating ratio and coverage ratio above 1.0. In fact, most bonds and loans require the utility to have a coverage ratio of at least 1.25. As state and federal funds for utility improvements have become much more difficult to obtain, these financial indicators have become more important for utilities. Being able to show and document the financial stability of the utility is an important part of getting funding for more capital improvements.

The operating ratio is perhaps the simplest measure of a utility's financial stability. In essence, the utility must be generating enough revenue to pay its operating expenses. The actual ratio is usually computed on a yearly basis, since many utilities may have monthly variations that do not reflect the overall performance. The total revenue is calculated by adding up all revenue generated by user fees, hook-up charges, taxes or assessments, interest income, and special income. Next determine the total operating expenses by adding up the

---

[5] Debt Service. The amount of money required annually to pay the (1) interest on outstanding debts; or (2) funds due on a maturing bonded debt or the redemption of bonds.

expenses of the utility, including administrative costs, salaries, benefits, energy costs, chemicals, supplies, fuel, equipment costs, equipment replacement fund, principal and interest payments, and other miscellaneous expenses.

EXAMPLE 1

The total revenues for a utility are $1,686,000 and the operating expenses for the utility are $1,278,899. The debt service expenses are $560,000. What is the operating ratio? What is the coverage ratio?

Known		Unknown
Total Revenue, $	= $1,686,000	Operating Ratio
Operating Expenses, $	= $1,278,899	Coverage Ratio
Debt Service Expenses, $	= $560,000	

1. Calculate operating ratio.

$$\text{Operating Ratio} = \frac{\text{Total Revenue, \$}}{\text{Operating Expenses, \$}}$$

$$= \frac{\$1,686,000}{\$1,278,899}$$

$$= 1.32$$

2. Calculate non-debt expenses.

$$\text{Non-Debt Expenses, \$} = \text{Operating Exp, \$} - \text{Debt Service Exp, \$}$$

$$= \$1,278,899 - \$560,000$$

$$= \$718,899$$

3. Calculate coverage ratio.

$$\text{Coverage Ratio} = \frac{\text{Total Revenue, \$} - \text{Non-Debt Expenses, \$}}{\text{Debt Service Expenses, \$}}$$

$$= \frac{\$1,686,000 - \$718,899}{\$560,000}$$

$$= 1.73$$

These calculations provide a good starting point for looking at the financial strength of the utility. Both of these calculations use the total revenue for the utility, which is an important component for any utility budgeting. As managers we often focus on the expense side and forget to look carefully at the revenue side of utility management. The fees collected by the utility, including hook-up fees and user fees, must accurately reflect the cost of providing service. These fees must be reviewed annually and they must be increased as expenses rise to maintain financial stability. Some other areas to examine on the revenue side include how often and how well user fees are collected, the number of delinquent accounts, and the accuracy of meters (for drinking water utilities). Some small communities have found they can cut their administrative costs significantly by switching to a quarterly billing cycle. The utility must have the support of the community to determine and collect user fees, and the utility must keep track of revenue generation as carefully as resource spending.

9.1 Budgeting

Budgeting for the utility is perhaps the most challenging task of the year for many managers. The list of needs usually is much larger than the possible revenue for the utility. The only way for the manager to prepare a good budget is to have good records from the year before. A system of recording or filing purchase orders or a requisition records system (see Section 13.5, "Procurement Records") must be in place to keep track of expenses and prevent spending money that is not in the budget.

To budget effectively, a manager needs to understand how the money has been spent over the last year, the needs of the utility, and how the needs should be prioritized. The manager also must take into account cost increases that cannot be controlled while trying to minimize the expenses as much as possible. The following problem is an example of the types of decisions a manager must make to keep the budget in line while also improving service from the utility.

EXAMPLE 2

A wastewater pump that has been in operation for 25 years pumps a constant 600 GPM through 47 feet of dynamic head. The pump uses 6,071 kilowatt-hours of electricity per month, at a cost of $0.085 per kilowatt-hr. The old pump efficiency has dropped to 63 percent. Assuming a new pump that operates at 86 percent efficiency is available for $9,730.00, how long would it take to pay for replacing the old pump?

Known		Unknown
Electricity, kW-hr/mo	= 6,071 kW-hr/mo	New Pump Payback Time, yr
Electricity Cost, $/kW-hr	= $0.085/kW-hr	
Old Pump Efficiency, %	= 63%	
New Pump Efficiency, %	= 86%	
New Pump Cost, $	= $9,730	

1. Calculate old pump operating costs in dollars per month.

$$\text{Old Pump Operating Costs, \$/mo} = (\text{Electricity, kW-hr/mo})(\text{Electricity Cost, \$/kW-hr})$$

$$= (6,071 \text{ kW-hr/mo})(\$0.085/\text{kW-hr})$$

$$= \$516.04/\text{mo}$$

2. Calculate new pump operating electricity requirements.

$$\text{New Pump Electricity, kW-hr/mo} = (\text{Old Pump Electricity, kW-hr/mo})\frac{(\text{Old Pump Eff, \%})}{(\text{New Pump Eff, \%})}$$

$$= (6,071 \text{ kW-hr/mo})\frac{(63\%)}{(86\%)}$$

$$= 4,447 \text{ kW-hr/mo}$$

3. Calculate new pump operating costs in dollars per month.

$$\text{New Pump Operating Costs, \$/mo} = (\text{Electricity, kW-hr/mo})(\text{Electricity Cost, \$/kW-hr})$$

$$= (4{,}447 \text{ kW-hr/mo})(\$0.085/\text{kW-hr})$$

$$= \$378.03/\text{mo}$$

4. Calculate annual cost savings of new pump.

$$\text{Cost Savings, \$/yr} = (\text{Old Costs, \$/mo} - \text{New Costs, \$/mo})(12 \text{ mo/yr})$$

$$= (\$516.04/\text{mo} - \$378.03/\text{mo})(12 \text{ mo/yr})$$

$$= \$1{,}656.12/\text{yr}$$

5. Calculate the new pump payback time in years.

$$\text{Payback Time, yr} = \frac{\text{Initial Cost, \$}}{\text{Savings, \$/yr}}$$

$$= \frac{\$9{,}730.00}{\$1{,}656.12/\text{yr}}$$

$$= 5.9 \text{ years}$$

In this example a payback time of 5.9 years is acceptable and would probably justify the expense for a new pump. This calculation was a simple payback calculation which did not take into account the maintenance on each pump, depreciation, and inflation. Many excellent references are available from EPA to help utility managers make more complex decisions about purchasing new equipment.

The annual report should be used to help develop the budget so that long-term planning will have its place in the budgeting process. The utility manager must track revenue generation and expenses with adequate records to budget effectively. The manager must also get input from other personnel in the utility as well as community leaders as the budgeting process proceeds. This input from others is invaluable to gain support for the budget and to keep the budget on track once adopted.

9.2 Equipment Repair/Replacement Fund

To adequately plan for the future, every utility must have a repair/replacement fund. The purpose of this fund is to generate additional revenue to pay for the repair and replacement of capital equipment as the equipment wears out. To prepare adequately for this repair/replacement, the manager should make a list of all capital equipment (this is called an asset inventory) and estimate the replacement cost for each item. The expected life span of the equipment must be used to determine how much money should be collected over time. When a treatment plant is new, the balance in repair/replacement fund should be increasing each year. As the plant gets older, the funds will have to be used and the balance may get dangerously low as equipment breakdowns occur. Perhaps the hardest job for the utility manager is to maintain a positive balance in this account with the understanding that this account is not meant to generate a "profit" for the utility but rather to plan for future equipment needs. In wastewater facilities constructed with federal funds under the construction grant program, providing an adequate repair/replacement fund was one of the grant conditions, but if this repair/replacement fund hasn't been reviewed annually, it must be updated.

To set up a repair/replacement fund for your utility, you should first put together a list of the equipment required for each process in your utility. Once you have this list, you need to estimate the life expectancy of the equipment and the replacement cost. From this list you can predict the amount of money you should set aside each year so that when each piece of equipment wears out, you will have enough money to replace that piece of equipment. Several EPA publications listed in Section 16, "Additional Reading," at the end of this manual are excellent references for utility planning.

9.3 Capital Improvements and Funding in the Future

A capital improvements fund must be a part of the utility budget and included in the operating ratio. Your responsibility as the utility manager is to be sure that everyone, your governing body and the public, understands the capital improvement fund is not a profit for the utility but a replacement fund to keep the utility operating in the future.

Capital planning starts with a look at changes in the community. Where are the areas of growth in the community, where are the areas of decline, and what are the anticipated changes in industry within the community? After identifying the changing needs in the community, you should examine the existing utility structure. Identify your weak spots (in the collection or distribution system, or with in-plant processes). Make a list of the areas that will be experiencing growth, weak spots in the system, and anticipated new regulatory requirements. The list should include expected capital improvements that will need to be made over the next year, two years, five years, and ten years. You can use the information in your annual reports and other operational logs to help compile the list.

Once you have compiled this information, prioritize the list and make a timetable for improving each of the areas. Starting at the top of the priority list, estimate the costs for improvements and incorporate these costs into your capital improvement budget. The calculations you have made previously, including operating ratio, coverage ratio, and payback time, will all be useful in prioritizing and streamlining your list of needs. Another useful ratio is the corrective to preventive maintenance ratio.

You may find that some of your capital improvement needs could be met in more than one way. How do you decide which of several options is most cost-effective? How do you compare fundamentally different solutions? For example, assume your community's population is growing rapidly and you will need to increase treated water production by 20 percent by the end of the next ten years. Your existing wells cannot provide that amount of additional water and your filtration equipment is operating at 90 percent of design capacity. Possible solutions might include a combination of the following options, some of which might be implemented immediately while others might be brought on line in five or ten years:

- Rehabilitate some declining wells,
- Drill additional new wells,
- Develop an available surface water source,
- Install another filtration unit, and/or
- Install additional distribution system storage reservoirs.

To compare alternative plans you will need to calculate the present value (or *PRESENT WORTH*[6]) of each plan; that is,

---

[6] *Present Worth. The value of a long-term project expressed in today's dollars. Present worth is calculated by converting (discounting) all future benefits and costs over the life of the project to a single economic value at the start of the project. Calculating the present worth of alternative projects makes it possible to compare them and select the one with the largest positive (beneficial) present worth or minimum present cost.*

the costs and benefits of each plan in today's dollars. This is done by identifying all the costs and benefits of each alternative plan over the same time period or time horizon. Costs should include not only the initial purchase price or construction costs, but also financing costs over the life of the loans or bonds and all operation and maintenance costs. Benefits include all of the revenue that would be produced by this facility or equipment, including connection and user fees. With the help of an experienced accountant, apply standard inflation, depreciation, and other economic discount factors to calculate the present value of all the benefits and costs of each plan during the same planning period. This will give you the cost of each plan in the equivalent of today's dollars.

Remember to involve all of your local officials and the public in development of this capital improvement budget so they understand what will be needed.

Long-term capital improvements such as a new plant or a new treatment process are usually anticipated in your 10-year or 20-year projection. These long-term capital improvements usually require some additional financing. The basic ways for a utility to finance capital improvements are through general obligation bonds, revenue bonds, or loan funding programs.

General obligation bonds or *ad valorem* (based on value) taxes are assessed based on property taxes. These bonds usually have a lower interest rate and longer payback time, but the total bond limit is determined for the entire community. This means that the water or wastewater utility will have available only a portion of the total bond capacity of the community. These bonds are not often used for funding water and wastewater utility improvements today.

The second type of bond, the revenue bond, is commonly used to fund utility improvements. This bond has no limit on the amount of funds available and the user charges provide repayment on the bond. To qualify for these bonds, the utility must show sound financial management and the ability to repay the bond. As the utility manager you should be aware of the provisions of the bond. Be sure the bond has a call date, which is the first date when you can pay off the bond. The common practice is for a 20-year bond to have a 10-year call date and for a 15-year bond to have an 8-year call date. The bond will also have a call premium, which is the amount of extra funds needed to pay off the debt on the call date. You should try to get your bonds a call premium of no more than 102 percent par. This means that for a debt of $200,000 on the call date, the total payoff would be $204,000, which includes the extra two percent for the call premium. You will need to get help from a financial advisor to prepare for and issue the bonds. These advisors will help you negotiate the best bond structure for your community.

Special assessment bonds may be used to extend services into specific areas. The direct users pay the capital costs and the assessment is usually based on frontage or area of real estate. These special assessments carry a greater risk to investors but may be the best way to extend service to some areas.

The most common way to finance water and wastewater improvements in the past has been federal and state grant programs. The Block Grants from HUD are still available for some projects and Rural Utilities Service (RUS) loans may also be used as a funding source. In addition, state revolving fund (SRF) programs provide loans (but not direct grants) for improvements. The SRF program has already been implemented with wastewater improvements and the new Safe Drinking Water Regulations include an SRF program for funding water treatment improvements. These SRF programs will be very competitive and utilities must provide evidence of sound financial management to qualify for these loans. You should contact your state regulatory agency to find out more about the SRF program in your state.

9.4 Financial Assistance

Many small utility systems need additional funds to repair and upgrade their systems. Potential funding sources include loans and grants from federal and state agencies, banks, foundations, and other sources. Some of the federal funding programs for small public utility systems include:

- Appalachian Regional Commission (ARC),
- Department of Housing and Urban Development (HUD) (provides Community Development Block Grants),
- Economic Development Administration (EDA),
- Indian Health Service (IHS), and
- Rural Utilities Service (RUS)(formerly Farmer's Home Administration (FmHA) and Rural Development Administration (RDA)).

For additional information regarding potential funding sources, contact the Water Environment Federation, www.wef.org or the American Water Works Association, www.awwa.org.

Another valuable contact is the Environmental Financing Information Network (EFIN) which provides information on financing alternatives for state and local environmental programs and projects in the form of abstracts of publications, case studies, and contacts. Contact U.S. Environmental Protection Agency (EPA), EFIN (mail code 2731R), Ariel Rios Building, 1200 Pennsylvania Avenue, NW, Washington, DC 20460. Phone (202) 564-4994 and FAX (202) 565-2587.

Also many states have one or more special financing mechanisms for small public utility systems. These funds may be in the form of grants, loans, bonds, or revolving loan funds. Contact your state for more information.

QUESTIONS

Write your answers in a notebook and then compare your answers with those on page 52.

9.0A List the three main areas of financial management for a utility.

9.0B How is a utility's operating ratio calculated?

9.3A Why is it important for a manager to consult with other utility personnel and with community leaders during the budget process?

9.3B How can long-term capital improvements be financed?

9.3C What is a revenue bond?

10 OPERATIONS AND MAINTENANCE

10.0 The Manager's Responsibilities

A utility manager's specific operation and maintenance (O&M) responsibilities vary depending on the size of the utility. At a small utility, the manager may oversee all utility operations while also serving as chief operator and supervising a small staff of operations and maintenance personnel. In larger utility agencies, the manager may have no direct, day-to-day responsibility for operations and maintenance but is ultimately

32 Utility Management

responsible for efficient, cost-effective operation of the entire utility. Whether large or small, every utility needs an effective operations and maintenance program.

10.1 Purpose of O&M Programs

The purpose of O&M programs is to maintain design functionality (capacity) and/or to restore the system components to their original condition and thus functionality. Stated another way, does the system perform as designed and intended? The ability to effectively operate and maintain a water or wastewater utility so it performs as intended depends greatly on proper design (including selection of appropriate materials and equipment), construction and inspection, acceptance, and system start-up. Permanent system deficiencies that affect O&M of the system are frequently the result of these phases. O&M staff should be involved at the beginning of each project, including planning, design, construction, acceptance, and start-up. When a utility system is designed with future O&M considerations in mind, the result is a more effective O&M program in terms of O&M cost and performance.

Effective O&M programs are based on knowing what components make up the system, where they are located, and the condition of the components. With that information, proactive maintenance can be planned and scheduled, rehabilitation needs identified, and long-term Capital Improvement Programs (CIPs) planned and budgeted. High-performing agencies have all developed performance measurements of their O&M program and track the information necessary to evaluate performance.

10.2 Types of Maintenance

Water or wastewater system maintenance can be either a proactive or a reactive activity. Commonly accepted types of maintenance include three classifications: corrective maintenance, preventive maintenance, and predictive maintenance.

Corrective maintenance, including emergency maintenance, is reactive. For example, a piece of equipment or a system is allowed to operate until it fails, with little or no scheduled maintenance occurring prior to the failure. Only when the equipment or system fails is maintenance performed. Reliance on reactive maintenance will always result in poor system performance, especially as the system ages. Utility agencies taking a corrective maintenance approach are characterized by:

- The inability to plan and schedule work,
- The inability to budget adequately,
- Poor use of resources, and
- A high incidence of equipment and system failures.

Emergency maintenance involves two types of emergencies: normal emergencies and extraordinary emergencies. Public utilities are faced with normal emergencies on a daily basis, whether it is a water main break or a blockage in a sewer. Normal emergencies can be reduced by an effective maintenance program. Extraordinary emergencies, such as high-intensity rainstorms, hurricanes, floods, and earthquakes, will always be unpredictable occurrences. However, the effects of extraordinary emergencies on the utility's performance can be minimized by implementation of a planned maintenance program and development of a comprehensive emergency response plan (see Section 11).

Preventive maintenance is proactive and is defined as a programmed, systematic approach to maintenance activities. This type of maintenance will always result in improved system performance except in the case where major chronic problems are the result of design and/or construction flaws that cannot be corrected by O&M activities. Proactive maintenance is performed on a periodic (preventive) basis or an as-needed (predictive) basis. Preventive maintenance can be scheduled on the basis of specific criteria such as equipment operating time since the last maintenance was performed, or passage of a certain amount of time (calendar period). For example, performing television inspection of a gravity sewer system on a five-year cycle (calendar period) would require that 20 percent of the system be televised each year. At the end of the five-year period, the cycle would start over again. Similarly, lubrication of motors is frequently based on running time.

The major elements of a good preventive maintenance program include the following:

- Planning and scheduling,
- Records management,
- Spare parts management,
- Cost and budget control,
- Emergency repair procedures, and
- Training program.

Some benefits of taking a preventive maintenance approach are:

- Maintenance can be planned and scheduled,
- Work backlog can be identified,
- Adequate resources necessary to support the maintenance program can be budgeted,
- Capital Improvement Program (CIP) items can be identified and budgeted for, and
- Human and material resources can be used effectively.

Predictive maintenance, which is also proactive, is a method of establishing baseline performance data, monitoring performance criteria over a period of time, and observing changes in performance so that failure can be predicted and maintenance can be performed on a planned, scheduled basis. Knowing the condition of the system makes it possible to plan and schedule maintenance on an "as required" basis and thus avoid unnecessary maintenance. An example of predictive maintenance in a collection system would be visually inspecting manholes or monitoring flows; when changes in flow conditions are observed, cleaning is scheduled accordingly.

In reality, every agency operates their system with corrective and emergency maintenance, preventive maintenance, and predictive maintenance methods. The goal, however, is to reduce the corrective and emergency maintenance efforts by performing preventive maintenance which will minimize system failures that result in stoppages and overflows.

System performance is frequently a reliable indicator of how the system is operated and maintained. Agencies that rely primarily on corrective maintenance as their method of operating and maintaining the system are never able to focus on preventive and predictive maintenance since most of their resources are directed at corrective maintenance activities and it is difficult to free up these resources to begin developing preventive maintenance programs. For an agency to develop an effective proactive maintenance program, they must add initial resources over and above those currently existing in order to establish preventive and predictive maintenance programs.

10.3 Benefits of Managing Maintenance

The goal of managing maintenance is to minimize investments of labor, materials, money, and equipment. In other words, we want to manage our human and material resources as effectively as possible, while delivering a high level of service to our customers. The benefits of an effective operation and maintenance program are as follows:

- Ensuring the availability of facilities and equipment as intended.

- Maintaining the reliability of the equipment and facilities as designed. Utility systems are required to operate 24 hours per day, 7 days per week, 365 days per year. Reliability is a critical component of the operation and maintenance program. If equipment and facilities are not reliable, then the ability of the system to perform as designed is impaired.

- Maintaining the value of the investment. Water and wastewater systems represent major capital investments for communities and are major capital assets of the community. If maintenance of the system is not managed, equipment and facilities will deteriorate through normal use and age. Maintaining the value of the capital asset is one of the utility manager's major responsibilities. Accomplishing this goal requires both ongoing investment to maintain existing facilities and equipment and extend the life of the system, and establishment of a comprehensive O&M program.

- Obtaining full use of the system throughout its design life.

- Collecting accurate information and data on which to base the operation and maintenance of the system and justify requests for the financial resources necessary to support it.

QUESTIONS

Write your answers in a notebook and then compare your answers with those on page 52.

10.1A What is the purpose of an operation and maintenance (O&M) program?

10.2A What are the three common types of maintenance?

10.2B List the major elements of a good preventive maintenance program.

11 EMERGENCY RESPONSE PLAN

Contingency planning is an essential facet of utility management and one that is often overlooked. Although utilities in various locations will be vulnerable to somewhat different kinds of natural disasters, the effects of these disasters in many cases will be quite similar. As a first step toward an effective contingency plan, each utility should make an assessment of its own vulnerability and then develop and implement a comprehensive plan of action.

All utilities suffer from common problems such as equipment breakdowns and leaking pipes. During the past few years there has also been an increasing amount of vandalism, civil disorder, toxic spills, and employee strikes which have threatened to disrupt utility operations. In observing today's international tension and the potential for nuclear war or the effects of terrorist-induced chemical or biological warfare, water utilities must seriously consider how to respond. Natural disasters such as floods, earthquakes, hurricanes, forest fires, avalanches, and blizzards are a more or less routine occurrence for some utilities. When such catastrophic emergencies occur, the utility must be prepared to minimize the effects of the event and have a plan for rapid recovery. Such preparation should be a specific obligation of every utility manager.

Start by assessing the vulnerability of the utility during various types of emergency situations. If the extent of damage can be estimated for a series of most probable events, the weak elements can be studied and protection and recovery operations can center on these elements. Although all elements are important for the utility to function, experience with disasters points out elements that are most subject to disruption. These elements are:

1. The absence of trained personnel to make critical decisions and carry out orders,

2. The loss of power to the utility's facilities,

3. An inadequate amount of supplies and materials, and

4. Inadequate communication equipment.

The following steps should be taken in assessing the vulnerability of a system:

1. Identify and describe the system components,

2. List assumed disaster characteristics,

3. Estimate disaster effects on system components,

4. Estimate customer demand for service following a potential disaster, and

5. Identify key system components that would be primarily responsible for system failure.

If the assessment shows a system is unable to meet estimated requirements because of the failure of one or more critical components, the vulnerable elements have been identified. Repeating this procedure using several "typical" disasters will usually point out system weaknesses. Frequently the same vulnerable element appears for a variety of assumed disaster events.

An emergency operations plan need not be too detailed, since all types of emergencies cannot be anticipated and a complex response program can be more confusing than helpful. Supervisory personnel must have a detailed description of their responsibilities during emergencies. They will need information, supplies, equipment, and the assistance of trained personnel. All these can be provided through a properly constructed emergency operations plan that is not extremely detailed.

The following outline can be used as the basis for developing an emergency operations plan:

1. Make a vulnerability assessment,
2. Inventory organizational personnel,
3. Provide for a recovery operation (plan),
4. Provide training programs for operators in carrying out the plan,
5. Coordinate with local and regional agencies such as the health, police, and fire departments to develop procedures for carrying out the plan,
6. Establish a communications procedure, and
7. Provide protection for personnel, plant equipment, records, and maps.

By following these steps, an emergency plan can be developed and maintained even though changes in personnel may occur. "Emergency Simulation" training sessions, including the use of standby power, equipment, and field test equipment will ensure that equipment and personnel are ready at times of emergency.

A list of phone numbers for operators to call in an emergency should be prepared and posted by a phone for emergency use. The list should include:

1. Plant supervisor,
2. Director of public works or head of utility agency,
3. Police,
4. Fire,
5. Doctor (2 or more),
6. Ambulance (2 or more),
7. Hospital (2 or more).

If appropriate for your utility, also include the following phone numbers on the emergency list:

8. Chlorine supplier and manufacturer,
9. CHEMTREC (800-424-9300) for the hazardous chemical spills; sponsored by the Manufacturing Chemists Association,
10. U.S. Coast Guard's National Response Center (800-424-8802),
11. Local and state poison control centers, and
12. Local hazardous materials spill response team.

You should prepare a list for your plant NOW, if you have not already done so, and update it annually when you update your emergency plan.

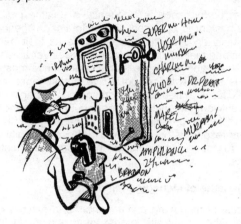

QUESTIONS

Write your answers in a notebook and then compare your answers with those on page 52.

11.0A What is the first step toward an effective contingency plan for emergencies?

11.0B Why is too detailed an emergency operations plan not needed or even desirable?

12 HOMELAND DEFENSE

World events in recent years have heightened concern in the United States over the security of the critical water and wastewater infrastructure. Water and wastewater pipelines form an extensive network that runs near or beneath key buildings and roads and is physically close to many communication and transportation networks. Significant damage to these utility facilities could result in: loss of life, catastrophic environmental damage to rivers, lakes, and wetlands, contamination of drinking water supplies, long-term public health impacts, destruction of fish and shellfish production, and disruption to commerce, the economy, and our normal way of life.

Water and wastewater facilities have been identified as targets for international and domestic terrorism. This knowledge, coupled with the responsibility of utilities to provide a safe and healthful workplace, requires that management establish rules to protect the workers as well as the facilities. Emergency action and fire prevention plans must identify what steps need to be taken when the threat analysis indicates a potential for attack. These plans must be in writing and be practiced periodically so that all workers know what actions to take.

Some actions that should be taken at all times to reduce the possibility of a terrorist attack are:

- Ensure that all visitors sign in and out of the facilities with a positive ID check,
- Reduce the number of visitors to a minimum,
- Discourage parking by the public near critical buildings to eliminate the chances of car bombs,
- Be cautious with suspicious packages that arrive,

- Be aware of the hazardous chemicals used and how to defend against spills,
- Keep emergency numbers posted near telephones and radios,
- Patrol the facilities frequently, looking for suspicious activity or behavior, and
- Maintain, inspect, and use your PPE (hard hats, respirators).

The following recommendations by the EPA[7] include many straightforward, common-sense actions a utility can take to increase security and reduce threats from terrorism.

Guarding Against Unplanned Physical Intrusion

- Lock all doors and set alarms at your office, pumping stations, treatment plants, and vaults, and make it a rule that doors are locked and alarms are set.
- Limit access to facilities and control access to pumping stations, chemical and fuel storage areas, giving close scrutiny to visitors and contractors.
- Post guards at treatment plants and post "Employee Only" signs in restricted areas.
- Secure hatches, metering vaults, manholes, and other access points to the sanitary collection system or the water distribution system.
- Increase lighting in parking lots, treatment bays, and other areas with limited staffing.
- Control access to computer networks and control systems and change the passwords frequently.
- Do not leave keys in equipment or vehicles at any time.

Making Security a Priority for Employees

- Conduct background security checks on employees at hiring and periodically thereafter.
- Develop a security program with written plans and train employees frequently.
- Ensure all employees are aware of established procedures for communicating with law enforcement, public health, environmental protection, and emergency response organization.
- Ensure that employees are fully aware of the importance of vigilance and the seriousness of breaches in security.
- Make note of unaccompanied strangers on the site and immediately notify designated security officers or local law enforcement agencies.
- If possible, consider varying the timing of operational procedures so that, to anyone watching for patterns, the pattern changes.
- Upon the dismissal of an employee, change pass codes and make sure keys and access cards are returned.
- Provide customer service staff with training and checklists of how to handle a threat if it is called in.

Coordinating Actions for Effective Emergency Response

- Review existing emergency response plans and ensure that they are current and relevant.
- Make sure employees have the necessary training in emergency operating procedures.
- Develop clear procedures and chains-of-command for reporting and responding to threats and for coordinating with emergency management agencies, law enforcement personnel, environmental and public health officials, consumers, and the media. Practice the emergency procedures regularly.
- Ensure that key utility personnel (both on and off duty) have access to critical telephone numbers and contact information at all times. Keep the call list up to date.
- Develop close relationships with local law enforcement agencies and make sure they know where critical assets are located. Ask them to add your facilities to their routine rounds.
- Work with local industries to ensure that their pretreatment facilities are secure.
- Report to county or state health officials any illness among the employees that might be associated with water or wastewater contamination.
- Immediately report criminal threats, suspicious behavior, or attacks on utility facilities to law enforcement officials and the nearest field office of the Federal Bureau of Investigation.

Investing in Security and Infrastructure Improvements

- Assess the vulnerability of the water distribution system, wastewater collection system, major pumping or lift stations, water and wastewater treatment plants, chemical and fuel storage areas, water intake pipes and wastewater outfall pipes, and other key infrastructure elements.
- Assess the vulnerability of the storm water collection system. Determine where large pipes run near or beneath government buildings, banks, commercial districts, industrial facilities, or are next to major communication and transportation networks. Move as quickly as possible with the most obvious and cost-effective physical improvements, such as perimeter fences, security lighting, and tamper-proofing manhole covers and valve boxes.
- Improve computer system and remote operational security.
- Use local citizen watches.
- Seek financing for more expensive and comprehensive system improvements.

The U.S. Terrorism Alert System (Figure 4) is a color-coded system that identifies the potential for terrorist activity and suggests specific actions to be taken. Your safety plan should identify the actions that your facility will take when the threat level changes. Tables 6 and 7 show examples of security measures that should be taken to improve safety at a utility facilities when the threat level is YELLOW and when it is ORANGE. (The utility's safety plan should include similar lists of actions for the RED, BLUE, and GREEN levels as well.)

---

[7] Adapted from "What Wastewater Utilities Can Do Now to Guard Against Terrorist and Security Threats," U.S. Environmental Protection Agency, Office of Wastewater Management, October 2001.

36 Utility Management

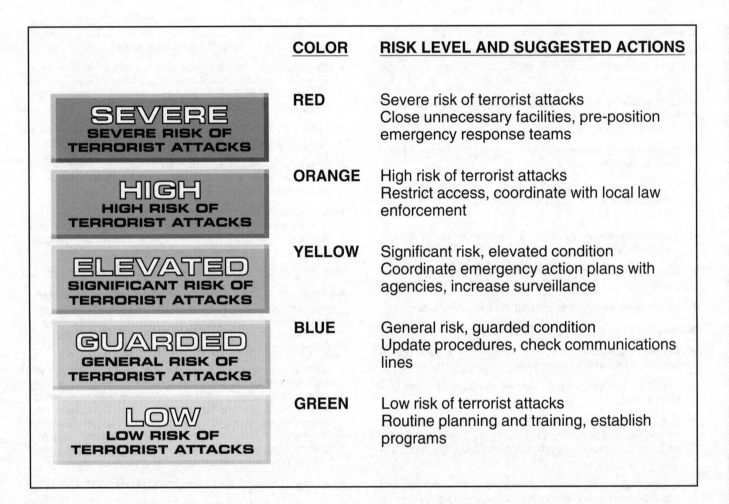

Fig. 4 Threat level categories established by the U.S. Department of Homeland Defense

TABLE 6 SECURITY MEASURES FOR THREAT LEVEL YELLOW (CONDITION ELEVATED)

Continue to introduce all measures listed in BLUE: Condition Guarded.	
Detection	**Prevention**
• To the extent possible, increase the frequency and extent of monitoring the flow coming into and leaving the treatment facility and review results against baseline quantities. Increase review of operational and analytical data (including customer complaints) with an eye toward detecting unusual variability (as an indicator of unexpected changes in the system). Variations due to natural or routine operational variability should be considered first. • Increase surveillance activities in wastewater collection and treatment facilities or water supply, treatment, and distribution facilities.	• Carefully review all facility tour requests before approving. If allowed, implement security measures to include list of names prior to tour, request identification of each attendee prior to tour, prohibit backpacks, duffle bags, and cameras, and identify parking restrictions. • On a daily basis, inspect the interior and exterior of buildings in regular use for suspicious activity or packages, signs of tampering, or indications of unauthorized entry. • Implement mail room security procedures. Follow guidance provided by the United States Postal Service.
Preparedness	**Protection**
• Continue to review, update and test emergency response procedures and communication protocols. • Establish unannounced security spot checks (such as verification of personal identification and door security) at access control points for critical facilities. • Increase frequency for posting employee reminders of the threat situation and about events that constitute security violations. • Ensure employees understand notification procedures in the event of a security breach. • Conduct security audit of physical security assets, such as fencing and lights, and repair or replace missing/broken assets. Remove debris from along fence lines that could be stacked to facilitate scaling. • Maximize physical control of all equipment and vehicles; make them inoperable when not in use (for example, lock steering wheels, secure keys, chain, and padlock on front-end loaders). • Review draft communications on potential incidents; brief media relations personnel of potential for press contact and/or issuance of press releases. • Ensure that list of sensitive customers (such as government agencies and industrial users) within the service area is accurate and shared with appropriate public health officials. • Contact neighboring water and wastewater utilities to review coordinated response plans and mutual aid during emergencies. • Review whether critical replacement parts are available and accessible. • Identify any work/project taking place in proximity to events where large attendance is anticipated. Consult with the event organizers and local law enforcement regarding contingency plans, security awareness, and site accessibility and control.	• Verify the identity of all persons entering the water or wastewater utility. Mandate visible use of identification badges. Randomly check identification badges and cards of those on the premises. • At the discretion of the facility manager or security director, remove all vehicles and objects (such as trash containers) located near mission critical facility security perimeters and other sensitive areas. • Verify the security of critical information systems (for example, Supervisory Control and Data Acquisition (SCADA), Internet, e-mail) and review safe computer and Internet access procedures with employees to prevent cyber intrusion. • Consider steps needed to control access to all areas under the jurisdiction of the utility. • Implement critical infrastructure facility surveillance and security plans. • At the beginning and end of each work shift, as well as at other regular and frequent intervals, inspect the interior and exterior of buildings in regular use for suspicious packages, persons, and circumstances. • Lock and regularly inspect all buildings, rooms, and storage areas not in regular use.

TABLE 7 SECURITY MEASURES FOR THREAT LEVEL ORANGE (CONDITION HIGH)

Continue to introduce all measures listed in YELLOW: Condition Elevated.	
Detection	**Prevention**
• Increase the frequency and extent of monitoring activities. Review results against baseline measurements. • Confirm that county and state health officials are on high alert and will inform the utility of any potential waterborne illnesses. • If a neighborhood watch-type program is in place, notify the community and request increased awareness.	• Discontinue tours and prohibit public access to all operational facilities. • Consider requesting increased law enforcement surveillance, particularly of critical assets and otherwise unprotected areas. • Limit access to computer facilities. No outside visitors. • Increase monitoring of computer and network intrusion detection systems and security monitoring systems.
Preparedness	**Protection**
• Confirm that emergency response and laboratory analytical support network are ready for deployment 24 hours per day, 7 days a week. • Reaffirm liaison with local police, intelligence, and security agencies to determine likelihood of an attack on the utility's personnel or facilities and consider appropriate protective measures (such as road closing and extra surveillance). • Practice communications procedures with local authorities and others cited in the facility's emergency response plan. • Post frequent reminders for staff and contractors of the threat level, along with a reminder of what events constitute security violations. • Ensure employees are fully aware of emergency response communication procedures and have access to contact information for relevant law enforcement, public health, environmental protection, and emergency response organizations. • Have alternative water supply plan ready to implement (for example, bottled water delivery for employees and other critical business uses). • Place all emergency management and specialized response teams on full alert status. • Ensure personal protective equipment (PPE) and specialized response equipment is checked, issued, and readily available for deployment. • Review all plans, procedures, guidelines, personnel details, and logistical requirements related to the introduction of a higher threat condition level.	• Evaluate the need to staff the water or wastewater treatment facility at all times. • Increase security patrol activity to the maximum level sustainable and ensure tight security in the vicinity of mission critical facilities. Vary the timing of security patrols. • Request employees change their passwords on critical information management systems. • Limit building access points to the absolute minimum, strictly enforce entry control procedures. Identify and protect all designated vulnerable points. Give special attention to vulnerable points outside of the critical facility. • Lock all exterior doors except the main facility entrance(s). Check all visitors' purpose, intent, and identification. Ensure that contractors have valid work orders. Require visitor's to sign in upon arrival; verify and record their identifying information. Escort visitors at all times when they are in the facility.

QUESTIONS

Write your answers in a notebook and then compare your answers with those on page 52.

12.0A What are the color codes and risk levels for the U. S. Terrorism Alert System?

12.0B What are some steps that can be taken to protect the utility's computer systems from a cyber attack?

13 SAFETY PROGRAM

The utility manager is responsible for the safety of the agency's personnel and the public exposed to the water or wastewater utility's operations. Therefore, the manager must develop and administer an effective safety program and must provide new employee safety training as well as ongoing training for all employees. The basic elements of a safety program include a safety policy statement, safety training and promotion, and accident investigation and reporting.

13.0 Policy Statement

A safety policy statement should be prepared by the top management of the utility. The purpose of the statement is to let employees know that the safety program has the full support of the agency and its management. The statement should:

1. Define the goals and objectives of the program,
2. Identify the persons responsible for each element of the program,
3. Affirm management's intent to enforce safety regulations, and
4. Describe the disciplinary actions that will be taken to enforce safe work practices.

Give a copy of the safety policy statement to every current employee and each new employee during orientation. Figure 5 is an example of a safety policy statement for a water supply utility.

13.1 Responsibilities

The following list of responsibilities for safety is from the *PLANT MANAGER'S HANDBOOK*.[8] These responsibilities represent a typical list but may be incomplete if your agency is subject to stricter local, state, and/or federal regulations than what is shown here. Check with your safety professional.

Management has the responsibility to:

1. Formulate a written safety policy,
2. Provide a safe workplace,
3. Set achievable safety goals,
4. Provide adequate training, and
5. Delegate authority to ensure that the program is properly implemented.

The manager is the key to any safety program. Implementation and enforcement of the program is the responsibility of the manager. The manager also has the responsibility to:

1. Ensure that all employees are trained and periodically retrained in proper safe work practices,
2. Ensure that proper safety practices are implemented and continued as long as the policy is in effect,
3. Investigate all accidents and injuries to determine their cause,
4. Institute corrective measures where unsafe conditions or work methods exist, and
5. Ensure that equipment, tools, and the work are maintained to comply with established safety standards.

The utility operators are the direct beneficiaries of a safety program. They share the responsibility to:

1. Observe prescribed work procedures with respect to personal safety and that of their co-workers,
2. Report any detected hazard to a manager immediately,
3. Report any accident, including a minor accident that causes minor injuries,
4. Report near-miss accidents so that hazards can be removed or procedures changed to avoid problems in the future, and
5. Use all protective devices and safety equipment supplied to reduce the possibility of injury.

13.2 Hazard Communication Program and Worker Right-To-Know (RTK) Laws

In the past few years there has been an increased emphasis nationally on hazardous materials and wastes. Much of this attention has focused on hazardous and toxic waste dumps and the efforts to clean them up after the long-term effects on human health were recognized. Each year thousands of new chemical compounds are produced for industrial, commercial, and household use. Frequently, the long-term effects of these chemicals are unknown. As a result, federal and state laws have been enacted to control all aspects of hazardous materials handling and use. These laws are more commonly known as Worker Right-To-Know (RTK) laws. The law that has had the greatest impact is the *OCCUPATIONAL SAFETY AND HEALTH ACT OF 1970 (OSHA)*,[9] Public Law 91-596, which took effect on December 29, 1970.

The intent of the OSHA regulations is to create a place of employment that is free from recognized hazards that could cause serious physical harm or death to an operator (or other employee). In many cases, the individual states have the authority under the OSHA Standard to develop their own state RTK laws and most states have adopted their own laws. The Federal OSHA Standard 29 *CFR*[10] 1910.1200—Hazard Communication forms the basis of most of these state RTK laws. Unfortunately, state laws vary significantly from state to state. The state laws that have been passed are at least as stringent

---

[8] *PLANT MANAGER'S HANDBOOK* (MOP SM-4), Water Environment Federation (WEF), no longer in print.

[9] OSHA (O-shuh). The Williams-Steiger **O**ccupational **S**afety and **H**ealth **A**ct of 1970 (OSHA) is a federal law designed to protect the health and safety of industrial workers, including the operators of water supply and treatment systems and wastewater treatment plants. The Act regulates the design, construction, operation, and maintenance of water supply systems, water treatment plants, wastewater collection systems, and wastewater treatment plants. OSHA also refers to the federal and state agencies which administer the OSHA regulations.

[10] Code of Federal Regulations (CFR). A publication of the United States Government which contains all of the proposed and finalized federal regulations, including environmental regulations.

SAFETY POLICY STATEMENT

It is the policy of the Las Vegas Valley Water District that every employee shall have a safe and healthy place to work. It is the District's responsibility; its greatest asset, the employees and their safety.

When a person enters the employ of the District, he or she has a right to expect to be provided a proper work environment, as well as proper equipment and tools, so that they will be able to devote their energies to the work without undue danger. Only under such circumstances can the association between employer and employee be mutually profitable and harmonious. It is the District's desire and intention to provide a safe workplace, safe equipment, proper materials, and to establish and insist on safe work methods and practices at all times. It is a basic responsibility of all District employees to make the SAFETY of human beings a matter for their daily and hourly concern. This responsibility must be accepted by everyone who works at the District, regardless of whether he or she functions in a management, supervisory, staff, or the operative capacity. Employees must use the SAFETY equipment provided; Rules of Conduct and SAFETY shall be observed; and SAFETY equipment must not be destroyed or abused. Further, it is the policy of the Water District to be concerned with the safety of the general public. Accordingly, District employees have the responsibility of performing their duties in such a manner that the public's safety will not be jeopardized.

The joint cooperation of employees and management in the implementation and continuing observance of this policy will provide safe working conditions and relatively accident-free performance to the mutual benefit of all involved. The Water District considers the SAFETY of its personnel to be of primary importance, and asks each employee's full cooperation in making this policy effective.

Fig. 5 Safety policy statement
(Permission of Las Vegas Valley Water District)

as the federal standard and, in most cases, are even more stringent. State laws are also under continuous revision and, because a strong emphasis is being placed on hazardous materials and worker exposure, state laws can be expected to be amended in the future to apply to virtually everybody in the workplace. Managers should become familiar with both the state and federal OSHA regulations that apply to their organizations.

The basic elements of a hazard communication program are described in the following paragraphs.

1. Identify Hazardous Materials. While there are thousands of chemical compounds which could be considered hazardous, focus on the ones to that operators and other personnel in your utility are most likely to be exposed.

2. Obtain Chemical Information and Define Hazardous Conditions. Once the inventory of hazardous chemicals is complete, the next step is to obtain specific information on each of the chemicals. This information is generally incorporated into a standard format form called the *MATERIAL SAFETY DATA SHEET (MSDS)*.[11] This information is commonly available from manufacturers. Many agencies request an MSDS when the purchase order is generated and will refuse to accept delivery of the shipment if the MSDS is not included. Figure 6 shows OSHA's standard MSDS form, but other forms are also acceptable provided they contain all of the required information.

   The purpose of the MSDS is to have a readily available reference document that includes complete information on common names, safe exposure level, effects of exposure, symptoms of exposure, flammability rating, type of first-aid procedures, and other information about each hazardous substance. Operators should be trained to read and understand the MSDS forms. The forms themselves should be stored in a convenient location where they are readily available for reference.

3. Properly Label Hazards. Once the physical, chemical, and health hazards have been identified and listed, a labeling and training program must be implemented. To meet labeling requirements on hazardous materials, specialized labeling is available from a number of sources, including commercial label manufacturers. Exemptions to labeling requirements do exist, so consult your local safety regulatory agency for specific details.

4. Train Operators. The last element in the hazard communication program is training and making information available to utility personnel. A common-sense approach eliminates the confusing issue of which of the thousands of

---

[11] MSDS. **M**aterial **S**afety **D**ata **S**heet. A document which provides pertinent information and a profile of a particular hazardous substance or mixture. An MSDS is normally developed by the manufacturer or formulator of the hazardous substance or mixture. The MSDS is required to be made available to employees and operators whenever there is the likelihood of the hazardous substance or mixture being introduced into the workplace. Some manufacturers are preparing MSDSs for products that are not considered to be hazardous to show that the product or substance is NOT hazardous.

Safety Program 41

Material Safety Data Sheet
May be used to comply with
OSHA's Hazard Communication Standard,
29 CFR 1910.1200 Standard must be
consulted for specific requirements.

U.S. Department of Labor
Occupational Safety and Health Administration
(Non-Mandatory Form)
Form Approved
OMB No. 1218-0072

IDENTITY *(As Used on Label and List)*

Note: *Blank spaces are not permitted. If any item is not applicable, or no information is available, the space must be marked to indicate that.*

Section I

Manufacturer's Name	Emergency Telephone Number
Address *(Number, Street, City, State, and ZIP Code)*	Telephone Number for Information
	Date Prepared
	Signature of Preparer *(optional)*

Section II — Hazardous Ingredients/Identity Information

Hazardous Components (Specific Chemical Identity: Common Name(s))	OSHA PEL	ACGIH TLV	Other Limits Recommended	% *(optional)*

Section III — Physical/Chemical Characteristics

Boiling Point		Specific Gravity ($H_2O = 1$)	
Vapor Pressure (mm Hg.)		Melting Point	
Vapor Density (Air = 1)		Evaporation Rate (Butyl Acetate = 1)	
Solubility in Water			
Appearance and Odor			

Section IV — Fire and Explosion Hazard Data

Flash Point (Method Used)	Flammable Limits	LEL	UEL
Extinguishing Media			
Special Fire Fighting Procedures			
Unusual Fire and Explosion Hazards			

(Reproduce locally)   OSHA 174, Sept. 1985

Fig. 6 Material Safety Data Sheet

Section V — Reactivity Data

Stability	Unstable		Conditions to Avoid
	Stable		

Incompatibility *(Materials to Avoid)*

Hazardous Decomposition or Byproducts

Hazardous Polymerization	May Occur		Condition to Avoid
	Will Not Occur		

Section VI — Health Hazard Data

Route(s) of Entry: Inhalation?   Skin?   Ingestion?

Health Hazards *(Acute and Chronic)*

Carcinogenicity   NTP?   IARC Monographs?   OSHA Regulated?

Signs and Symptoms of Exposure

Medical Conditions
Generally Aggravated by Exposure

Emergency and First Aid Procedures

Section VII — Precautions for Safe Handling and Use

Steps to Be Taken in Case Material is Released or Spilled

Waste Disposal Method

Precautions to be Taken in Handling and Storing

Other Precautions

Section VIII — Control Measures

Respiratory Protection (Specify Type)

Ventilation	Local Exhaust	Special
	Mechanical *(General)*	Other

Protective Gloves	Eye Protection

Other Protective Clothing or Equipment

Work/Hygienic Practices

Page 2   * USGPO 1986-491-529/45775

Fig. 6 Material Safety Data Sheet (continued)

substances operators should be trained for, and concentrates on those that they will be exposed to or use in everyday maintenance routines.

The Hazard Communication Standard and the individual state requirements are obviously a very complex set of regulations. Remember, however, the ultimate goal of these regulations is to provide additional operator protection. These standards and regulations, once the intent is understood, are relatively easy to implement.

13.3 Confined Space Entry Procedures

CONFINED SPACES[12, 13] pose significant risks for a large number of workers, including many utility operators. OSHA has therefore defined very specific procedures to protect the health and safety of operators whose jobs require them to enter or work in a confined space. The regulations (which can be found in the Code of Federal Regulations at 29 CFR 1910.146) require conditions in the confined space to be tested and evaluated before anyone enters the space. If conditions exceed OSHA's limits for safe exposure, additional safety precautions must be taken and a confined space entry permit (Figure 7) must be approved by the appropriate authorities prior to anyone entering the space.

The managers of water and wastewater utilities may or may not be involved in the day-to-day details of enforcing the agency's confined space policy and procedures. However, every utility manager should be aware of the current OSHA requirements and should ensure that the utility's policies not only comply with current regulations, but that the agency's policies are vigorously enforced for the safety of all operators.

13.4 Reporting

Regardless of the size of the utility, recordkeeping is an important part of an effective safety program. All injuries should be reported, even if they are minor in nature, so as to establish a record in case the injury develops into a serious injury. It may be difficult at a later date to prove whether the accident occurred on or off the job and this information may determine who is responsible for the costs. The responsibility for reporting accidents affects several levels of personnel. First, of course, is the injured person. Next, it is the responsibility of the supervisor, and finally, it is the

Responsibility of Management to review the causes and take steps to prevent such accidents from happening in the future.

Accident report forms may be very simple. However, they must record all details required by law and all data needed for statistical purposes. The forms shown in Figures 8 and 9 are examples for you to consider for use in your utility.

In addition to reports needed by the utility, other reports may be required by state or federal agencies. For example, vehicle accident reports must be submitted to local police departments. If a member of the public is injured, additional forms are needed because of possible subsequent claims for damages. If the accident is one of occupational injury, causing lost time, other reports may be required. Follow-up investigations to identify causes and responsibility may require the development of other specific types of record forms.

Emphasis on the prevention of future accidents cannot be overstressed. We must identify the causes of accidents and implement whatever measures are necessary to protect operators from becoming injured.

S SAFETY FIRST
A ACCIDENTS COST LIVES
F FASTER IS NOT ALWAYS BETTER
E EXPECT THE UNEXPECTED
T THINK BEFORE YOU ACT
Y YOU CAN MAKE THE DIFFERENCE

ACCIDENTS DON'T JUST HAPPEN... THEY ARE CAUSED!

QUESTIONS

Write your answers in a notebook and then compare your answers with those on page 52.

13.0A Who is responsible for the implementation of a safety program?

13.0B What should be included in a utility's policy statement on safety?

13.2A List the basic elements of a hazard communication program.

13.3A What are a manager's responsibilities for ensuring the safety of operators entering or working in confined spaces?

13.4A Why should a report be prepared whenever an injury occurs?

---

[12] Confined Space. Confined space means a space that:
  A. Is large enough and so configured that an employee can bodily enter and perform assigned work; and
  B. Has limited or restricted means for entry or exit (for example, manholes, tanks, vessels, silos, storage bins, hoppers, vaults, and pits are spaces that may have limited means of entry); and
  C. Is not designed for continuous employee occupancy.
  (Definition from the Code of Federal Regulations (CFR) Title 29 Part 1910.146.)

[13] Confined Space, Permit-Required (Permit Space). A confined space that has one or more of the following characteristics:
  • Contains or has a potential to contain a hazardous atmosphere,
  • Contains a material that has the potential for engulfing an entrant,
  • Has an internal configuration such that an entrant could be trapped or asphyxiated by inwardly converging walls or by a floor which slopes downward and tapers to a smaller cross section, or
  • Contains any other recognized serious safety or health hazard.
  (Definition from the Code of Federal Regulations (CFR) Title 29 Part 1910.146.)

Confined Space Pre-Entry Checklist/Confined Space Entry Permit

Date and Time Issued: _____ Date and Time Expires: _____ Job Site/Space I.D.: _____

Job Supervisor: _____ Equipment to be worked on: _____ Work to be performed: _____

Standby personnel: _____ _____ _____

1. Atmospheric Checks: Time _____ Oxygen _____ % Toxic _____ ppm
   Explosive _____ % LEL Carbon Monoxide _____ ppm

2. Tester's signature: _____

3. Source isolation: (No Entry)   N/A   Yes   No
   Pumps or lines blinded,
   disconnected, or blocked   ( )   ( )   ( )

4. Ventilation Modification: N/A   Yes   No
   Mechanical   ( )   ( )   ( )
   Natural ventilation only   ( )   ( )   ( )

5. Atmospheric check after isolation and ventilation: Time _____
   Oxygen _____ % > 19.5% < 23.5%   Toxic _____ ppm < 10 ppm H_2S
   Explosive _____ % LEL < 10%   Carbon Monoxide _____ ppm < 35 ppm CO

Tester's signature: _____

6. Communication procedures: _____
7. Rescue procedures: _____
_____

8. Entry, standby, and backup persons   Yes   No
   Successfully completed required training?   ( ) ( )
   Is training current?   ( ) ( )

9. Equipment: N/A   Yes   No
   Direct reading gas monitor tested   ( ) ( ) ( )
   Safety harnesses and lifelines for entry and standby persons   ( ) ( ) ( )
   Hoisting equipment   ( ) ( ) ( )
   Powered communications   ( ) ( ) ( )
   SCBAs for entry and standby persons   ( ) ( ) ( )
   Protective clothing   ( ) ( ) ( )
   All electric equipment listed for Class I, Division I,
   Groups A, B, C, and D, and nonsparking tools   ( ) ( ) ( )

10. Periodic atmospheric tests:
    Oxygen: ____% Time ___; ____% Time ___; ____% Time ___; ____% Time ___;
    Explosive: ____% Time ___; ____% Time ___; ____% Time ___; ____% Time ___;
    Toxic: ____ppm Time ___; ____ppm Time ___; ____ppm Time ___; ____ppm Time ___;
    Carbon Monoxide: ____ppm Time ___; ____ppm Time ___; ____ppm Time ___; ____ppm Time ___;

We have reviewed the work authorized by this permit and the information contained herein. Written instructions and safety procedures have been received and are understood. Entry cannot be approved if any brackets ( ) are marked in the "No" column. This permit is not valid unless all appropriate items are completed.

Permit Prepared By: (Supervisor) _____ Approved By: (Unit Supervisor) _____

Reviewed By: (CS Operations Personnel) _____
                    (Entrant)   (Attendant)   (Entry Supervisor)

This permit to be kept at job site. Return job site copy to Safety Office following job completion.

Fig. 7   Confined space pre-entry checklist/confined space entry permit

Safety Program 45

Date _____

Name of injured employee _____ Employee # _____ Area _____
Date of accident _____ Time _____ Employee's Occupation _____
Location of accident _____ Nature of injury _____
Name of doctor _____ Address _____
Name of hospital _____ Address _____
Witnesses (name & address) _____
_____

PHYSICAL CAUSES

Indicate below by an "X" whether, in your opinion, the accident was caused by:

_____	Improper guarding	_____	No mechanical cause
_____	Defective substances or equipment	_____	Working methods
_____	Hazardous arrangement	_____	Lack of knowledge or skill
_____	Improper illumination	_____	Wrong attitude
_____	Improper dress or apparel	_____	Physical defect
_____	Not listed (describe briefly) _____		

UNSAFE ACTS

Sometimes the injured person is not directly associated with the causes of an accident. Using an "X" to represent the injured worker and an "O" to represent any other person involved, indicate whether, in your opinion, the accident was caused by:

_____	Operating without authority	_____	Unsafe loading, placement & etc.
_____	Failure to secure or warn	_____	Took unsafe position
_____	Working at unsafe speed	_____	Worked on moving equipment
_____	Made safety device inoperative	_____	Teased, abused, distracted & etc.
_____	Unsafe equipment or hands instead of equipment	_____	Did not use safe clothing or personal protective equipment
_____	No unsafe act		
_____	Not listed (describe briefly) _____		

_____

What job was the employee doing? _____
What specific action caused the accident? _____
What steps will be taken to prevent recurrence? _____
_____
_____

Date of Report _____ Immediate Supervisor _____

REVIEWING AUTHORITY

Comments: | Comments:

Safety Officer _____ Date | Department Director _____ Date

Fig. 8 Supervisor's accident report

46 Utility Management

INJURED: COMPLETE THIS SECTION

Name _____ Age _____ Sex _____

Address _____

Title _____ Dept. Assigned _____

Place of Accident _____

Street or Intersection _____

Date _____ Hour _____ A.M. _____ P.M. _____

Type of Job You Were Doing When Injured
_____

Object Which Directly Injured You   Part of Body Injured
_____

How Did Accident Happen? (Be specific and give details; use back of sheet if necessary.)

_____

	Yes	No
Did You Report Accident or Exposure at Once? (Explain "No")	☐	☐
Did You Report Accident or Exposure to Supervisor? Give Name	☐	☐
Were There Witnesses to Accident or Exposure? Give Names	☐	☐
Did You See a Doctor? (If Yes, Give Name)	☐	☐
Are You Going to See a Doctor? (Give Name)	☐	☐

_____   _____
           Date   Signature

SUPERVISOR: COMPLETE THIS SECTION — (Return to Personnel as Soon as Possible)

	Yes	No
Was an Investigation of Unsafe Conditions and/or Unsafe Acts Made? If Yes, Please Submit Copy.	☐	☐
Was Injured Intoxicated or Behaving Inappropriately at Time of Accident? (Explain "Yes")	☐	☐

Date Disability   Last Day   Date Back
Commenced _____   Wages Earned _____   on Job _____

Date Report Completed _____ 20 ____   Signed By _____

                                                                    Title _____

Distribution: Canary - Department Head, Pink - Supervisor, White - Personnel

Fig. 9 Accident report

14 RECORDKEEPING

14.0 Purpose of Records

Accurate records are essential for effective utility management and to satisfy legal requirements. Records are also a valuable source of information. They can save time when trouble develops and provide proof that problems were identified and solved. Pertinent and complete records should be used as a basis for plant operation, interpreting results of water or wastewater treatment, preparing preventive maintenance programs, and preparing budget requests. When accurately kept, records provide a sound basis for design of future changes or expansions of the treatment plant. If legal questions or problems occur in connection with the treatment processes or the operation of the plant, accurate and complete records will provide evidence of what actually occurred and what procedures were followed.

14.1 Computer Recordkeeping Systems

Until fairly recently, water supply system recordkeeping has been done manually. The current availability of low-cost personal computer systems puts automation of many manual bookkeeping functions within the means of all water utilities.

To automate your recordkeeping functions as they relate to customer billing, you will need to develop a simple database management system that will create tables similar to those illustrated in this chapter. This can be readily accomplished by use of standard spreadsheet software programs which are available in the marketplace at a cost of $300 to $400. Hardware including a personal computer, data storage system, and a printer can be purchased for under $5,000.

Excellent computer software packages are being developed and offered to assist utility managers. SURF (Small Utility Rates and Finances) has been developed by the American Water Works Association (AWWA). SURF is a self-guided, interactive spreadsheet application designed to assist small drinking water systems in developing budgets, setting user rates, and tracking expenses. SURF requires very little computer or software knowledge and can be used by system operators, bookkeepers, and managers to improve the financial management practices of their utilities. SURF can print out three separate modules: (1) system budget, (2) user rate(s), and (3) system expenses.

SURF hardware and software requirements are modest.

Hardware Requirements: IBM Compatible PC with at least 4 MB RAM (8 MB recommended). Color monitor. Ink jet or laser jet printer.

Software Requirements: DOS (Version 5 or newer), Windows 95 or higher and Microsoft Excel 97.

The SURF software and an excellent user's manual are available free from the American Water Works Association. They can be obtained by calling the AWWA Small Systems Program at (800) 366-0107 or by downloading the program from the AWWA website (www.AWWA.org).

Two other computer programs available for water and wastewater utility managers for rate setting, impact fees, and financial planning are RateMod Pro ($1,495) and RateMod XP ($1,795). For details and assistance in selecting the most appropriate program for your utility, contact RateMod Associates at (202) 237-2455.

14.2 Types of Records

Many different types of records are required for effective management and operation of water supply, treatment, and distribution system facilities or wastewater collection, treatment, and disposal facilities. The following Sections, 14.3 through 14.7, describe some of the most important types of records that should be kept and Section 14.8 discusses how long records need to be kept.

14.3 Equipment and Maintenance Records

A good plant maintenance effort depends heavily on good recordkeeping. You will need to keep accurate records to monitor the operation and maintenance of each piece of plant equipment. Equipment control cards and work orders can be used to:

- Record important equipment data such as make, model, serial number, and date purchased,
- Record maintenance and repair work performed to date,
- Anticipate preventive maintenance needs, and
- Schedule future maintenance work.

Whenever a piece of equipment is changed, repaired, or tested, the work performed should be recorded on an equipment history card of some type. Complete, up-to-date equipment records will enable the plant operators to evaluate the reliability of equipment and will provide the basis for a realistic preventive maintenance program.

14.4 Plant Operations Data

Plant operations logs can be as different as the treatment plants whose information they record. The differences in amount, nature, and format of data are so significant that any attempt to prepare a "typical" log would be very difficult. For detailed information and example recordkeeping forms for a water supply utility, see *WATER TREATMENT PLANT OPERATION*, Volume I, Chapter 10, "Plant Operation," Section 10.6, "Operating Records and Reports." For further information regarding the types of daily and monthly operating records required for a wastewater treatment plant, see *OPERATION OF WASTEWATER TREATMENT PLANTS*, Volume II, Chapter 19, "Records and Report Writing," Section 19.11, "Type of Records."

14.5 Procurement Records

Ordering repair parts and supplies usually is done when the on-hand quantity of a stocked part or chemical falls below the reorder point, a new item is added to stock, or an item has been requested that is not stocked. Most organizations require employees to submit a requisition (Figure 10) when they need to purchase equipment or supplies. When the requisition has been approved by the authorized person (a supervisor or purchasing agent, in most cases) the items are ordered using a form called a purchase order. A purchase order contains a number of important items, including: (1) the date, (2) a complete description of each item and quantity needed, (3) prices, (4) the name of the vendor, and (5) a purchase order number.

A copy of the purchase order should be retained in a suspense file or on a clipboard until the ordered items arrive. This procedure helps keep track of the items that have been ordered but have not yet been received.

All supplies should be processed through the storeroom immediately upon arrival. When an item is received, it should be so recorded on an inventory card. The inventory card will keep

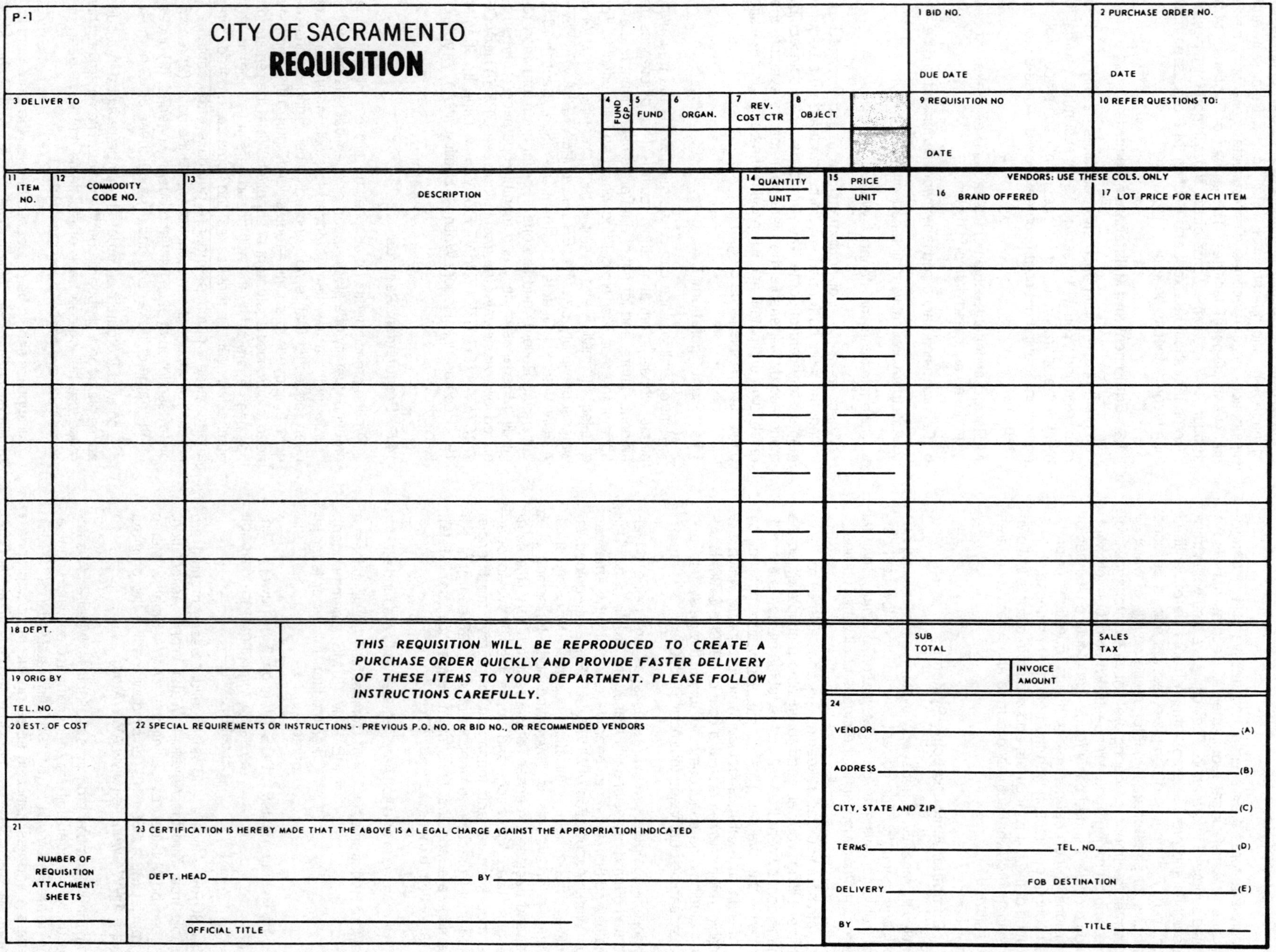

P-1

CITY OF SACRAMENTO
REQUISITION

1 BID NO.
DUE DATE
2 PURCHASE ORDER NO.
DATE

3 DELIVER TO
4 FUND GP.
5 FUND
6 ORGAN.
7 REV. COST CTR
8 OBJECT
9 REQUISITION NO
DATE
10 REFER QUESTIONS TO:

11 ITEM NO.	12 COMMODITY CODE NO.	13 DESCRIPTION	14 QUANTITY UNIT	15 PRICE UNIT	16 BRAND OFFERED	17 LOT PRICE FOR EACH ITEM

VENDORS: USE THESE COLS. ONLY

SUB TOTAL
SALES TAX
INVOICE AMOUNT

18 DEPT.
19 ORIG BY
TEL. NO.

THIS REQUISITION WILL BE REPRODUCED TO CREATE A PURCHASE ORDER QUICKLY AND PROVIDE FASTER DELIVERY OF THESE ITEMS TO YOUR DEPARTMENT. PLEASE FOLLOW INSTRUCTIONS CAREFULLY.

20 EST. OF COST
22 SPECIAL REQUIREMENTS OR INSTRUCTIONS - PREVIOUS P.O. NO. OR BID NO., OR RECOMMENDED VENDORS

21
NUMBER OF REQUISITION ATTACHMENT SHEETS

23 CERTIFICATION IS HEREBY MADE THAT THE ABOVE IS A LEGAL CHARGE AGAINST THE APPROPRIATION INDICATED
DEPT. HEAD ______ BY ______
OFFICIAL TITLE ______

24
VENDOR ______ (A)
ADDRESS ______ (B)
CITY, STATE AND ZIP ______ (C)
TERMS ______ TEL. NO. ______ (D)
DELIVERY ______ FOB DESTINATION (E)
BY ______ TITLE ______

Fig. 10 Requisition/purchase order form

track of the numbers of an item in stock, when last ordered, cost, and other information. Furthermore, by always logging in supplies immediately upon receipt, you are in a position to reject defective or damaged shipments and control shortages or errors in billing. Many utilities now use personal computers to keep track of orders and deliveries.

14.6 Inventory Records

An inventory consists of the supplies the treatment plant needs to keep on hand to operate the facility. These maintenance supplies may include repair parts, spare valves, electrical supplies, tools, and lubricants. The purpose of maintaining an inventory is to provide needed parts and supplies quickly, thereby reducing equipment downtime and work delays.

In deciding what supplies to stock, keep in mind the economics involved in buying and stocking an item as opposed to depending on outside availability to provide needed supplies. Is the item critical to continued plant or process operation? Should certain frequently used repair parts be kept on hand? Does the item have a shelf life?

Inventory costs can be held to a minimum by keeping on hand only those parts and supplies for which a definite need exists or which would take too long to obtain from an outside vendor. A "definite need" for an item is usually demonstrated by a history of regular use. Some items may be infrequently used but may be vital in the event of an emergency; these items should also be stocked. Take care to exclude any parts and supplies that may become obsolete, and do not stock parts for equipment scheduled for replacement.

14.7 Personnel Records

Documentation of all aspects of personnel management provides an important measure of legal protection for the utility. If an employee files a lawsuit alleging discriminatory hiring practices or treatment, harassment, breach of contract, or other grievance, the utility's ability to defend its practices and procedures will depend almost entirely on complete and accurate records. Similarly, if an employee is injured on the job, written records can help establish whether the utility was responsible for the accident.

Each personnel action should be fully documented in writing and filed. Even verbal discussions of a supervisor with an employee about job performance should be summarized in writing upon completion of the conversation. Also file copies of all written warnings and disciplinary actions.

An employee's personnel file should also contain a complete record of accomplishments, certificates earned, commendations received, and formal performance reviews.

As previously mentioned, personnel records often contain sensitive, confidential information; therefore, access to these records should be closely controlled.

14.8 Disposition of Plant and System Records

An important question is how long records should be kept. As a general rule, records should be kept as long as they may be useful or as long as legally required. Some information will become useless after a short time, while other data may be valuable for many years. Data that might be used for future design or expansion should be kept indefinitely. Laboratory data will always be useful and should be kept indefinitely. Regulatory agencies may require you to keep certain water quality analyses (bacteriological test results) and customer complaint records on file for specified time periods (10 years for chemical analyses and bacteriological tests).

Even if old records are not consulted every day, this does not lessen their potential value. For orderly records handling and storage, set up a schedule to periodically review old records and to dispose of those records that are no longer needed. A decision can be made when a record is established regarding the time period for which it must be retained.

QUESTIONS

Write your answers in a notebook and then compare your answers with those on page 53.

14.0A   What are some of the benefits of keeping complete, up-to-date records?

14.4A   List the important items usually contained on a purchase order.

14.5A   What is the purpose of maintaining an inventory?

14.7A   As a general rule, how long should utility records be kept?

15 ACKNOWLEDGMENTS

During the writing of this manual, Lynne Scarpa, Phil Scott, Chris Smith, and Rich von Langen, all members of California Water Environment Association (CWEA), provided many excellent materials and suggestions for improvement. Their generous contributions are greatly appreciated.

16 ADDITIONAL READING

1. *WATER UTILITY MANAGEMENT* (M5). Obtain from American Water Works Association (AWWA), Bookstore, 6666 West Quincy Avenue, Denver, CO 80235. Order No. 30005. ISBN 0-89867-063-2. Price to members, $68.50; nonmembers, $98.50; price includes cost of shipping and handling.

2. *A WATER AND WASTEWATER MANAGER'S GUIDE FOR STAYING FINANCIALLY HEALTHY*, U.S. Environmental Protection Agency. Obtain from National Technical Information Service (NTIS), 5285 Port Royal Road, Springfield, VA 22161. Order No. PB90-114455. EPA No. 430-9-89-004. Price, $33.50, plus $5.00 shipping and handling per order.

3. *WASTEWATER UTILITY RECORDKEEPING, REPORTING AND MANAGEMENT INFORMATION SYSTEMS*, U.S. Environmental Protection Agency. Obtain from National Technical Information Service (NTIS), 5285 Port Royal Road, Springfield, VA 22161. Order No. PB83-109348. EPA No. 430-9-82-006. Price, $39.50, plus $5.00 shipping and handling per order.

4. *SUPERVISION: CONCEPTS AND PRACTICES OF MANAGEMENT*, Ninth Edition, 2004, Hilgert, Raymond L., and Edwin Leonard, Jr. Obtain from Thomson Learning, Customer Service, 10650 Toebben Drive, Independence, KY 41051. ISBN 0-324-17881-6. Price, $92.95, plus shipping and handling.

**END OF LESSON 2 OF 2 LESSONS
ON
UTILITY MANAGEMENT**

Please answer the discussion and review questions next.

UTILITY MANAGEMENT
DISCUSSION AND REVIEW QUESTIONS
(Lesson 2 of 2 Lessons)

Write the answers to these questions in your notebook. The question numbering continues from Lesson 1.

18. With whom do managers need to communicate?
19. What information should be included in the utility's annual report?
20. List four steps that can be taken during a meeting to make sure it is a productive meeting.
21. What happens any time you or a member of your utility comes in contact with the public?
22. What attitude should management try to develop among its employees regarding the consumer?
23. What is the value of consumer complaints?
24. How do you measure financial stability for a utility?
25. How can a manager prepare a good budget?
26. What can happen when agencies rely primarily on corrective maintenance to keep the system running?
27. What types of information should be included in an emergency operations plan?
28. What is the intent of the OSHA regulations?
29. List four major types of records a utility must maintain.
30. Why is it important to document all aspects of personnel management?

UTILITY MANAGEMENT
SUGGESTED ANSWERS

ANSWERS TO QUESTIONS IN LESSON 1

Answers to questions on page 1.

1.0A Local utility demands on a utility manager include protection from environmental disasters with a minimum investment of money.

1.0B Changes in the environmental workplace are created by changes in the workforce and advances in technology.

Answers to questions on page 3.

2.0A The functions of a utility manager include planning, organizing, staffing, directing, and controlling.

2.0B In small communities the community depends on the manager to handle everything.

Answers to questions on page 6.

3.0A Utility planning must include operational personnel, local officials (decision makers), and the public.

4.0A The purpose of an organizational plan is to show who reports to whom and to identify the lines of authority.

4.0B Effective delegation is uncomfortable for many managers since it requires giving up power and responsibility. Many managers believe that they can do the job better than others, they believe that other employees are not well trained or experienced, and they are afraid of mistakes. The utility manager retains some responsibility even after delegating to another employee and, therefore, the manager is often reluctant to delegate or may delegate the responsibility but not the authority to get the job done.

4.0C   An important and often overlooked part of delegation is follow-up by the supervisor.

Answers to questions on page 8.

5.0A   Staffing responsibilities include hiring new employees, training employees, and evaluating job performance.

5.0B   The two personnel management concepts a manager should always keep in mind are "job-related" and "documentation."

5.1A   The steps involved in a staffing analysis include:

1. List the tasks to be performed,
2. Estimate the number of staff hours per year required to perform each task,
3. List the utility's current employees,
4. Assign tasks based on each employee's skills and abilities, and
5. Adjust the work assignments as necessary to achieve the best possible fit between the work to be done and the personnel/skills available to do it.

5.2A   A qualifications profile is a clear statement of the knowledge, skills, and abilities a person must possess to perform the essential job duties of a particular position.

Answers to questions on page 11.

5.3A   The purpose of a job interview is to gain additional information about the applicants so that the most qualified person can be selected to fill a job opening.

5.3B   Protected groups include minorities, women, disabled persons, persons over 40 years of age, and union members.

5.3C   The "OUCH" principle stands for:

Objectivity,
Uniform treatment of employees,
Consistency, and
Having job relatedness.

5.4A   A new employee's safety training should begin on the first day of employment or as soon thereafter as possible.

Answers to questions on page 13.

5.5A   The purpose of a probationary period for new employees is to provide a time during which both the employer and employee can assess the "fit" between the job and the person.

5.5B   The compensation an employee receives for the work performed includes satisfaction, recognition, security, appropriate pay, and benefits.

5.5C   Utility managers should provide training opportunities for employees so they can keep informed of new technologies and regulations. Training for supervisors is also important to ensure that supervisors have the knowledge, skills, and attitude that will enable them to be effective supervisors.

Answers to questions on page 17.

5.5D   An employee's immediate supervisor should conduct the employee's performance evaluation.

5.5E   Dealing with employee discipline requires tact and skill. The manager or supervisor should stay flexible, calm, and open-minded.

5.5F   Common warning signs of potential violence include abusive language, threatening or confrontational behavior, assault, and brandishing a weapon.

Answers to questions on page 21.

5.5G   Harassment is any behavior that is offensive, annoying, or humiliating to an individual and that interferes with a person's ability to do a job. This behavior is uninvited, often repeated, and creates an uncomfortable or even hostile environment in the workplace.

5.5H   Types of behavior that could be considered sexual harassment include:

- Unwanted hugging, patting, kissing, brushing up against someone's body, or other inappropriate sexual touching,
- Subtle or open pressure for sexual activity,
- Persistent sexually explicit or sexist statements, jokes, or stories,
- Repeated leering or staring at a person's body,
- Suggestive or obscene notes or phone calls, and/or
- Display of sexually explicit pictures or cartoons.

5.5I   The best way to prevent harassment is to set an example by your own behavior and to keep communication open between employees. A manager must also be aware of and take action to prevent any type of harassment in the workplace.

5.5J   In general, the ADA defines disability as a physical or mental impairment that substantially limits one or more of the major life activities of an individual.

Answers to questions on page 22.

5.6A   The shop steward is elected by the union employees and is their official representative to management and the local union.

5.6B   Union contracts do not change the supervisor's delegated authority or responsibility. Operators must carry out the supervisor's orders and get the work done properly, safely, and within a reasonable amount of time. However, a contract gives a union the right to protest or challenge a supervisor's decision.

ANSWERS TO QUESTIONS IN LESSON 2

Answers to questions on page 25.

6.0A   A manager needs both written and oral communication skills.

6.0B   The most common written documents that a utility manager must write include memos, business letters, press releases, monitoring reports, monthly reports, and the annual report.

6.0C   The annual report should be a review of what and how the utility operated during the past year and also the goals for the next year.

Answers to questions on page 26.

7.0A   A utility manager may be asked to conduct meetings with employees, the governing board, the public, and with other professionals in your field.

7.0B   Before a meeting (1) prepare an agenda and distribute, (2) find an adequate meeting room, and (3) set a beginning and ending time.

52 Utility Management

Answers to questions on page 28.

8.0A The first step in organizing a public relations campaign is to establish objectives so you will have a clear idea of what you expect to achieve.

8.1A Employees can be informed about the utility's plans, practices, and goals through newsletters, bulletin boards, and regular, open communication between supervisors and subordinates.

8.2A Newspapers give more thorough, in-depth coverage to stories than do the broadcast media.

8.4A Practice is the key to effective public speaking.

8.6A Complaints can be a valuable asset in determining consumer acceptance and pinpointing problems. Customer calls are frequently the first indication that something may be wrong. Responding to complaints and inquiries promptly can save the utility money and staff resources, and minimize customer inconvenience.

Answers to questions on page 31.

9.0A The three main areas of financial management for a utility include providing financial stability for the utility, careful budgeting, and providing capital improvement funds for future utility expansion.

9.0B The operating ratio for a utility is calculated by dividing total revenues by total operating expenses.

9.3A It is important for a manager to get input from other personnel in the utility as well as community leaders as the budgeting process proceeds in order to gain support for the budget and to keep the budget on track once adopted.

9.3B The basic ways for a utility to finance long-term capital improvements are through general obligation bonds, revenue bonds, or loan funding programs.

9.3C A revenue bond is a debt incurred by the community, often to finance utility improvements. User charges provide repayment on the bond.

Answers to questions on page 33.

10.1A The purpose of an O&M program is to maintain design functionality and/or to restore the system components to their original condition and thus functionality. That is, to ensure that the system performs as designed and intended.

10.2A Commonly accepted types of maintenance include corrective maintenance, preventive maintenance, and predictive maintenance.

10.2B The major elements of a good preventive maintenance program include the following:

- Planning and scheduling,
- Records management,
- Spare parts management,
- Cost and budget control,
- Emergency repair procedures, and
- Training program.

Answers to questions on page 34.

11.0A The first step toward an effective contingency plan for emergencies is to make an assessment of vulnerability. Then a comprehensive plan of action can be developed and implemented.

11.0B A detailed emergency operations plan is not needed since all types of emergencies cannot be anticipated and a complex response program can be more confusing than helpful.

Answers to questions on page 39.

12.0A

COLOR	RISK LEVEL AND SUGGESTED ACTIONS
RED	Severe risk of terrorist attacks
ORANGE	High risk of terrorist attacks
YELLOW	Significant risk, elevated condition
BLUE	General risk, guarded condition
GREEN	Low risk of terrorist attacks

12.0B To protect a utility's computer systems from a cyber attack, review safe procedures with employees, limit access to computer facilities (no outside visitors), and ask employees to change their passwords frequently on critical information management systems.

Answers to questions on page 43.

13.0A The utility manager is responsible for the safety of the agency's personnel and the public exposed to the water or wastewater utility's operations. Therefore, the manager must develop and administer an effective safety program and must provide new employee safety training as well as ongoing training for all employees.

13.0B A utility's policy statement on safety should include:

1. A definition of the goals and objectives of the program,
2. Identification of the persons responsible for each element of the program,
3. A statement affirming management's intent to enforce safety regulations, and
4. A description of the disciplinary actions that will be taken to enforce safe work practices.

13.2A The basic elements of a hazard communication program include the following:

1. Identify hazardous materials,
2. Obtain chemical information and define hazardous conditions,
3. Properly label hazards, and
4. Train operators.

13.3A A utility manager may or may not be involved in the day-to-day details of enforcing the agency's confined space policy and procedures. However, every utility manager should be aware of the current OSHA requirements and should ensure that the utility's policies not only comply with current regulations, but that the agency's policies are vigorously enforced for the safety of all operators.

13.4A All injuries should be reported, even if they are minor in nature, so as to establish a record in case the injury develops into a serious injury. It may be difficult at a later date to prove whether the accident occurred on or off the job and this information may determine who is responsible for the costs.

Answers to questions on page 49.

14.0A Keeping complete, up-to-date records contributes to more effective utility management, helps to satisfy legal requirements, provides valuable operations and maintenance information, and assists in preparing budget requests. Accurate records provide a sound basis for design of future changes or expansions of the treatment plant. If legal questions or problems occur in connection with the treatment processes or the operation of the plant, accurate and complete records will provide evidence of what actually occurred and what procedures were followed.

14.4A A purchase order usually contains: (1) the date, (2) a complete description of each item and quantity needed, (3) prices, (4) the name of the vendor, and (5) a purchase order number.

14.5A The purpose of maintaining an inventory is to provide needed parts and supplies quickly, thereby reducing equipment downtime and work delays.

14.7A As a general rule, utility records should be kept for as long as they may be useful or as long as legally required.

UTILITY MANAGEMENT
FINAL EXAMINATION

This final examination was prepared TO HELP YOU review the material in the manual. The questions have been divided into the following types:

1. True-false,
2. Best answer,
3. Multiple choice, and
4. Short answers.

To work this examination:

1. Write the answer to each question in your notebook,
2. After you have worked a group of questions (you decide how many), check your answers with the suggested answers at the end of this exam, and
3. If you missed a question and don't understand why, reread the material in the manual.

You may wish to use this examination for review purposes when preparing for civil service and certification examinations.

Since you have already completed this course, you do not have to send your answers to California State University, Sacramento.

True-False

1. The workforce in the environmental field is fairly stable.
   1. True
   2. False

2. Advances in technology are creating changes in the environmental workplace.
   1. True
   2. False

3. An operator can serve two or more supervisors.
   1. True
   2. False

4. Delegation requires giving up power and responsibility.
   1. True
   2. False

5. Nothing will destroy morale more quickly than equal treatment of employees.
   1. True
   2. False

6. An emergency operations plan should be as detailed as possible.
   1. True
   2. False

Best Answer (Select only the closest or best answer.)

1. Which problem makes planning most difficult in smaller communities?
   1. Decline in population
   2. Lack of revenue
   3. Stringent regulations
   4. Untrained operators

2. What does authority mean?
   1. Authority means an act in which power is given to another person
   2. Authority means answering to those above in the chain of command
   3. Authority means being accountable for results
   4. Authority means the power and resources to do a job

3. What is the main purpose of a probationary period for new employees?
   1. Allow time for operator to become certified
   2. Assess the "fit" between the job and the person
   3. Inform employee of utility policies and procedures
   4. Provide safety training

4. Who should conduct an employee's performance evaluation?
   1. Employee of equal rank
   2. Employee's immediate supervisor
   3. Shop steward
   4. Utility manager

5. What is the best way to start a disciplinary meeting with an employee?
   1. Explaining that you want a response that is acceptable to everyone
   2. Identifying the problem
   3. Indicating you are looking for a positive solution
   4. Making a positive comment about the employee

6. What is the first step for developing a public relations plan?
   1. Contact news media
   2. Establish objectives
   3. Identify target audience
   4. Prepare promotional material

7. What must a utility manager have to prepare a good budget?
   1. A method of disregarding uncontrolled increases in costs
   2. A system that prevents spending money not in budget
   3. Good records from the previous year
   4. Sufficient revenue

56 Utility Management

8. What is the main intent of the OSHA regulations?
   1. Create a safe work environment
   2. Identify hazards in the workplace
   3. Protect the public from hazardous chemicals
   4. Train operators in safe procedures

9. What should a manager look at as a first step in capital planning?
   1. Changes in the community
   2. Current sources of revenue
   3. Existing utility structure
   4. New regulatory requirements

Multiple Choice (Select all correct answers.)

1. What kinds of demands are utility managers expected to meet?
   1. Compliance with regulations
   2. Perform with a minimum investment of money
   3. Population forecasts
   4. Preparation of utility for future
   5. Protection of environment

2. What problems will a utility face without adequate planning?
   1. Adequate staff
   2. Budget surpluses
   3. Inability to meet compliance regulations
   4. Inadequate service capacity
   5. System failures

3. What are the purposes of a utility organizational plan?
   1. Control activities of operators
   2. Identify the lines of authority
   3. Organize plant activities
   4. Provide a means for emergency planning
   5. Show who reports to whom

4. What steps should a supervisor take to conduct a performance evaluation?
   1. Ask employee to sign completed evaluation form
   2. Compare employee's performance against others' performance
   3. Fill out evaluation form
   4. Schedule general staff meeting
   5. Set performance goals for next year

5. Which of the following topics are acceptable on an employee evaluation form?
   1. Attendance
   2. Cooperation and relationships
   3. Initiative
   4. Job knowledge
   5. Quality of work

6. What kinds of behavior are considered sexual harassment? Behavior of a sexual nature that is
   1. Annoying
   2. Hostile
   3. Humiliating
   4. Offensive
   5. Uninvited

7. What are warning signs that an employee might become violent?
   1. Abusive language
   2. Brandishing a weapon
   3. Confrontational behavior
   4. Tardiness
   5. Threatening behavior

8. What should be done before a successful meeting?
   1. Find an adequate meeting room
   2. Involve all participants
   3. Prepare an agenda and distribute to participants
   4. Restate your understanding of results
   5. Set a beginning and ending time

9. What are some guidelines for a successful news interview?
   1. Ask to approve a draft of the story
   2. Give as much detail as possible
   3. If you don't know an answer, bluff
   4. Know your facts
   5. State key points early in the interview

10. What are a manager's responsibilities for financial management of a utility?
    1. Careful budgeting
    2. Collecting unpaid bills
    3. Providing capital improvement funds
    4. Providing financial stability
    5. Stopping services for delinquent accounts

11. What is debt service?
    1. Debts owed contractors for work completed
    2. Funds due on a maturing bonded debt
    3. Funds due on redemption of bonds
    4. Interest on outstanding debts
    5. Unpaid debts owed by customers

12. Effective O & M (operation and maintenance) programs are based on what types of information?
    1. Condition of system components
    2. Location of system components
    3. Replacement costs of components
    4. Types of components that make up the system
    5. Vulnerability of components in a disaster

13. What are the major elements of a good preventive maintenance program?
    1. Cost and budget control
    2. Large work backlog
    3. Planning and scheduling
    4. Spare parts management
    5. Training program

14. What are the basic elements of a safety program?
    1. Accident investigation
    2. Insurance coverage
    3. Recordkeeping
    4. Safety training for operators
    5. Written safety policy

15. What are the basic elements of a hazard communication program?
    1. Confined space entry procedures
    2. Identify hazardous materials
    3. Obtain chemical information and define hazardous conditions
    4. Properly label hazards
    5. Train operators

Short Answers

1. What two concepts should a manager keep in mind to avoid violating the rights of an employee or job applicant?
2. What information should be provided to a new employee during orientation?
3. What is a manager's responsibility for preventing harassment in the workplace?
4. What is the purpose of a repair/replacement fund?
5. What happens any time you or a member of your utility comes in contact with the public?
6. How do you measure financial stability for a utility?
7. What is the intent of the OSHA regulations?

UTILITY MANAGEMENT
SUGGESTED ANSWERS FOR FINAL EXAMINATION

True-False

1. False — The workforce in the environmental field is changing.
2. True — Advances in technology are creating changes in the environmental workplace.
3. False — An operator can serve only one supervisor.
4. True — Delegation requires giving up power and responsibility.
5. False — Morale will be destroyed by UNEQUAL treatment of employees.
6. False — Too much detail in an emergency operations plan is confusing.

Best Answer

1. 1 — A decline in population makes planning most difficult in smaller communities.
2. 4 — Authority means the power and resources to do a job.
3. 2 — The main purpose of a probationary period for new employees is to assess the "fit" between the job and the person.
4. 2 — An employee's immediate supervisor should conduct the employee's performance evaluation.
5. 4 — The best way to start a disciplinary meeting with an employee is by making a positive comment about the employee.
6. 2 — The first step for developing a public relations plan is to establish objectives.
7. 3 — In order to prepare a good budget, a utility manager must have good records from the previous year.
8. 1 — The main intent of the OSHA regulations is to create a safe workplace environment.
9. 1 — As a first step in capital planning, a manager should look at any changes in the community.

58 Utility Management

Multiple Choice

1. 1, 2, 4, 5 — Utility managers are expected to be in compliance with regulations, to perform with a minimum investment of money, to prepare their utility for the future, and to protect the environment.

2. 3, 4, 5 — Without adequate planning, a utility can expect to face the following problems: an inability to meet compliance regulations; an inadequate service capacity; and system failures.

3. 2, 5 — The purposes of a utility's organizational plan are to identify the lines of authority and to show who reports to whom.

4. 1, 3, 5 — In conducting a performance evaluation, a supervisor fills out an evaluation form, sets performance goals for the coming year, and asks the employee to sign the completed evaluation form.

5. 1, 2, 3, 4, 5 — Acceptable topics on an employee evaluation form include the following: attendance; cooperation and relationships; initiative; job knowledge; and quality of work.

6. 1, 2, 3, 4, 5 — Sexual harassment consists of behavior of a sexual nature that is annoying, hostile, humiliating, offensive, and uninvited.

7. 1, 2, 3, 5 — Warning signs that an employee might become violent include abusive language, brandishing a weapon, confrontational behavior, and threatening behavior.

8. 1, 3, 5 — To ensure a successful meeting, a few tasks must be performed ahead of time. These include finding an adequate meeting room, preparing an agenda, and setting a beginning and ending time.

9. 4, 5 — One guideline for a successful news interview is to know the facts. Also, state the key points early in the interview and provide details, if asked.

10. 1, 3, 4 — A utility manager's financial management responsibilities include careful budgeting, providing a capital improvement fund, and providing financial stability.

11. 2, 3, 4 — Debt service is a term that describes the funds due on a maturing bonded debt, the funds due on the redemption of bonds, and the interest on outstanding debts.

12. 1, 2, 4 — Effective O & M (operation and maintenance) programs are based on knowing the types, the location, and the condition of components that make up the system.

13. 1, 3, 4, 5 — The major elements of a good preventive maintenance program include cost and budget control, planning and scheduling, spare parts management, and training programs.

14. 1, 3, 4, 5 — The basic elements of a safety program include accident investigation, recordkeeping, operator safety training, and a written safety policy.

15. 2, 3, 4, 5 — A hazard communication program includes the following basic elements: identifying hazardous materials; obtaining chemical information and defining hazardous conditions; properly labeling hazards; and training operators.

Short Answers

1. To avoid violating the rights of an employee or job applicant, a manager should keep in mind two concepts: *job-related* and *documentation*. Any personnel action taken must be job-related, and all personnel actions must be properly documented.

2. During orientation on the first day of work, a new employee should be given all the information available in written and verbal form on the policies and practices of the utility including compensation, benefits, attendance expectations, alcohol and drug testing (if the utility does this), and employer/employee relations.

3. A manager's responsibility for preventing harassment in the workplace includes *being aware of* and *taking action to prevent* any type of harassment from occurring.

4. The purpose of a repair/replacement fund is to generate additional revenue that can be used to fund the repair and replacement of capital equipment as the equipment wears out.

5. Any time you or a member of your utility comes in contact with the public, you will have an impact on the quality of your public image.

6. Financial stability is measured by calculating (1) the operating ratio, and (2) the coverage ratio. Both of these ratios should be greater than 1.0.

7. The intent of the OSHA regulations is to create a place of employment that is free from recognized hazards that could cause serious physical harm or death to an operator (or other employee).

UTILITY MANAGEMENT
SUBJECT INDEX

A

ABC (Association of Boards of Certification), 12
ADA (Americans With Disabilities Act), 20
Accident report form, 43, 45, 46
Accountable, 3
Ad valorem taxes, 31
Advertising, job openings, 8
Affirmative action, 8
Americans With Disabilities Act (ADA), 20
Annual report, 25, 30
Applications, job, 8
Assessment bonds, 31
Assessment of system vulnerability, 33
Audience, communications, 23, 24, 25, 26, 27
Audiovisuals, 24
Authority, 3

B

Benefits, employee, 9, 12
Block Grants, 31
Bonds, 31
Budgeting, 29

C

Call date, 31
Capital improvements, 30
Capital planning, 30
Civil rights, 8
Communication, 23-25
Compensation, 12
Complaint procedure, 13
Computer recordkeeping systems, 47
Conducting meetings, 25
Confined space entry permit, 44
Confined space entry procedure, 43
Conservation, water, 26
Consumer inquiries, 27
Contract negotiations, 21
Controlling, 3
Corrective maintenance, 32
Coverage ratio, 28, 29, 30
Customer complaints, 27
Customer inquiries, 27

D

Debt service, 28
Delegation, 3
Directing, 1
Disciplinary problems, 13
Discriminatory hiring, 7, 8

Disposition of records, 49
Documentation, 6

E

Emergency maintenance, 32
Emergency response plan, 3, 33
Employee benefits, 9, 12
Employee evaluation, 13, 14
Employee orientation, 11
Employee problems, 13, 21
Employee-management relationship, 21
Employees, 7
Employer policies, 11, 17
Equal Employment Opportunity Act, 20
Equipment records, 47
Equipment repair/replacement fund, 30
Evaluation, performance, 13, 14

F

Family and Medical Leave Act, 20
Fees, user, 28, 29
Financial assistance, 31
Financial management, 28-31
Financial stability, 28
FmHA loans, 31
Functions of manager, 1
Fund, repair/replacement, 30

G

General obligation bonds, 30, 31
Goals, utility, 3
Grant programs, 31
Grievances, 13, 21
Groups, protected, 11

H

Harassment, 17, 18
Hazard communication program, 39
Hiring procedures, 6-11
Homeland defense, 34

I

Improvements, capital, 30
Improving system management, 2
Inquiries, pre-employment, 9
Interview, job, 8, 9
Interview, media, 26

J

Job
  duties, 1, 5
  interview, 8, 9

K

(NO LISTINGS)

L

Labor laws, 20, 21, 22
Listening skills, 23
Loan funding programs, 30, 31
Long-term capital improvements, 30, 31

M

MSDS (Material Safety Data Sheet), 40-42
Maintenance records, 47
Maintenance, types, 32
Management questions, 2
Management, utility, 1
Manager, 1
Manager's responsibilities, 1, 6, 31
Material Safety Data Sheet (MSDS), 40-42
Media, news, 26
Meetings, conducting, 25

N

National Labor Relations Act, 22
Negotiations, contract, 21
New employee orientation, 11
News media, 26
Number of employees needed, 7

O

OSHA (Occupational Safety and Health Act), 39
OUCH principle, 11
Operating ratio, 28, 29, 30
Operations and maintenance (O & M), 31, 32, 33
Oral communication, 23
Organizational plan, 3, 4
Organizing, 1, 3
Orientation, new employee, 11

P

Payback time, 30
Performance evaluation, 13, 14
Permit, confined space entry, 44
Personnel records, 6, 21, 49
Plan, organizational, 3, 4
Planning, 1, 3
Plant operations data, 47
Plant tours, 28
Population decline, 3
Predictive maintenance, 32
Pre-employment inquiries, 9
Present worth, 30
Preventive maintenance, 32
Probationary period, 11
Problem employees, 13

Procurement records, 47
Productive meetings, 25
Profile, qualifications, 7
Property taxes, 31
Protected groups, 11
Public relations, 26-28
Public speaking, 27
Purchase orders, 29, 47, 48

Q

Qualifications profile, 7
Questions
  interview, 9
  management, 2

R

Recordkeeping
  computer systems, 47
  disposition of records, 49
  equipment, 47
  inventory, 49
  maintenance, 47
  personnel, 6, 21, 49
  plant operations data, 47
  purchase order, 29, 47, 48
  purpose, 47
  requisition, 29, 47, 48
  types of records, 47
Records, 6
Repair/replacement fund, 30
Replacement cost, 30
Report writing, 23-25
Reporting, safety, 43
Requisition system, 29, 47, 48
Responsibility, 3
Retaliation, 20
Revenue bonds, 30, 31
Rural Utilities Service (RUS), 31

S

SRF (State Revolving Fund), 31
SURF (Small Utility Rates and Finances), 47
Safety
  accident report form, 43, 45, 46
  confined space entry procedures, 43, 44
  hazard communication program, 39, 40
  Material Safety Data Sheet (MSDS), 40
  policy statement, 39, 40
  program, 39-43
  reporting, 43
  responsibilities, 39
  training, 11
  Worker Right-To-Know (RTK) Laws, 39
Selection, employees, 8
Sexual harassment, 17
Shop steward, 21
Special assessment bonds, 31
Staffing
  advertising positions, 8
  Americans With Disabilities Act, 20
  applications, job, 8
  certification, 12
  compensation, 12
  disciplinary problems, 13

Staffing (continued)
    employee orientation, 11
    employee problems, 13, 21
    employee-management relationship, 21
    employees, 7
    employer policies, 11, 17
    Equal Employment Opportunity Act, 20
    evaluation, performance, 13, 14
    Family and Medical Leave Act, 20
    grievances, 13, 21
    groups, protected, 11
    harassment, 17, 18
    hiring procedures, 6-11
    interview questions, 9
    interviewing, 8
    labor laws, 20, 22
    manager's responsibilities, 1, 6
    National Labor Relations Act, 22
    new employee orientation, 11
    number of employees, 7
    orientation, new employees, 11
    paper screening, 8
    performance evaluation, 13, 14
    personnel records, 21
    policies, employer, 11, 17
    probationary period, 11
    problem employees, 13
    profile, qualifications, 7
    protected groups, 11
    qualifications profile, 7
    retaliation, 20
    selection process, 8, 11
    sexual harassment, 17, 18
    training, 12
    unions, 21
State Revolving Fund (SRF), 31
Steward, shop, 21

T

Target audience, 23, 24, 25, 26, 27
Telephone contacts, 27
Threat level categories, 36
Tours, plant, 28
Training, 12

U

U.S. Terrorism Alert System, 35
Unaccounted for water, 2
Unions, 21
Unity of command, 3
Utility management, 1

V

Violence, 16
Vulnerability assessment, 33

W

Water conservation, 26
Worker Right-To-Know (RTK) Laws, 39
Working conditions, 9
Written communication, 23-25

X, Y, and Z

(NO LISTINGS)

NOTES

NOTES

NOTES

NOTES

NOTES